507
CHA

Chahrour, Janet.

Zap! blink! taste! think! :
exciting life science for
HARPER J.H.S. LIBRARY
202571

Zap! Blink! Taste! Think!

Exciting Life Science for Curious Minds

Janet Parks Chahrour

Illustrated by
Abe Gurvin

F. Harper JHS
4000 Covell Blvd.
Davis, CA 95616

Dedication

To curious kids. They ask good questions and take time to think.

To teachers. They have generous hearts.

Text © copyright 2003 by Janet Parks Chahrour
Illustrations © copyright 2003 by Barron's Educational Series, Inc.

All rights reserved.
No part of this book may be reproduced in any form, by photostat, microfilm, xerography, or any other means, or incorporated into any information retrieval system, electronic or mechanical, without the written permission of the copyright owner.

Record sheets may be copied for classroom use without permission.

All inquiries should be addressed to:
Barron's Educational Series, Inc.
250 Wireless Boulevard
Hauppauge, New York 11788
http://www.barronseduc.com

Library of Congress Catalog Card No.: 2002028036

International Standard Book No.: 0-7641-1912-5

Library of Congress Cataloging-in-Publication Data
Chahrour, Janet
   Zap! blink! taste! think! : exciting life science for curious minds / Janet Parks Chahrour ; illustrated by Abe Gurvin.
      p. cm.
      Summary: Presents the procedures and concepts involved in twenty-four science experiments that can be done at home with readily available materials, exploring psychology, food chemistry, crime-solving, and horoscopes.
      Includes bibliographical references and index.
      Contents: Your senses—Food science—Create your own—Inspections and dissections—Psychology and beliefs.
      ISBN 0-7641-1912-5
      1. Psychology—Experiments—Juvenile literature.
   2. Psychology, Experimental—Juvenile literature.
   [1. Psychology—Experiments. 2. Science—Experiments.
   3. Experiments 4. Science projects.] I. Gurvin, Abe, ill. II. Title.
BF198.7 C43   2003
507'.8—dc21
                              2002028036

PRINTED IN HONG KONG
9 8 7 6 5 4 3 2 1

Acknowledgments

These activity ideas and approaches were woven together over twenty-five years with strands from diverse sources, including student discoveries and my own invention. I am grateful to many dynamic teachers, writers, and students who have sparked my curiosity through conversations, workshops, journals, books, and web sites.

Notable resources are:

Bottle Biology: An Idea Book for Exploring the World through Plastic Bottles and Other Recyclable Materials, by Mrill Ingram, Kendall/Hunt Publishing Co., 1993.

Cookwise; The Hows and Whys of Successful Cooking, by Shirley O. Corriher, William Morrow and Company, Inc., 1997.

COSI Science Museum, Columbus, Ohio.

The Exploratorium Web Site. The Museum of Science, Art, and Human Perception. Available at http://www.exploratorium.edu.

Fingerprinting, Great Explorations in Math and Science (Gems) Program, Lawrence Hall of Science, 1987.

Gardening Wizardry for Kids, by L. Patricia Kite, Barron's Educational Series, Inc., 1995.

The Popcorn Board, pamphlets and web site
http://www.popcorn.org
401 North Michigan Avenue
Chicago, IL 60611-1083

Pride and a Daily Marathon, by Jonathan Cole, The MIT Press, 1995.

Science Experiments You Can Eat, by Vicki Cobb, Scholastic Inc., 1972.

Secrets of the Psychics, NOVA video with James Randi, produced for PBS by the Science Unit at WGBH Boston, 1974.

Simple Kitchen Experiments; Learning Science with Everyday Foods, by Muriel Mandell, Sterling Publishing Co., Inc., 1994.

I am indebted to Ryan Chahrour, Lois Parks, Russell Parks, and the DeMaioribus family, Nick, Derek, Alex, Natalie, and Carmel, for their thorough reviews of the manuscript and many helpful suggestions. My appreciation also goes to Keith Holden, Martha Crotty, Jerry Judge, Jane Biddinger, Judy Roush, Chris Specht, Andrew Butler, and Sam Tumolo for their valuable feedback.

Contents

What's So Special About This Book? / 1
How to Use This Book / 2
A Note to Parents / 4
A Note to Teachers / 5
Thinking Like a Scientist / 7
Master List of Materials / 8

ACTIVITIES AND RECORD SHEETS / 9

Your Senses
1. Mirror Mirror / 10
2. Flavor Detective / 16
3. What You See Is What You Get . . . Or Is It? / 22
4. Depth Perception / 28
5. Where Did I Put My Toes? / 35

Food Science
6. Food Preservation I: Dried Apples and Fruit Roll-Ups / 42
7. Food Preservation II: Refrigerator Pickles / 49
8. Enzyme Excitement in Garlic / 57
9. Testing Foods for Fat and Starch / 64
10. Snap! Crackle! Popcorn! / 68

Create Your Own
11. Boogie Woogie Butter / 76
12. Yogurt, Please / 83
13. Sprout Jungle / 89
14. Herb Garden / 95
15. Kitchen Compost / 103
16. Balance for Small Weights / 109

Inspections and Dissections
17. Water in Ketchup? / 114
18. Seed Survival / 120
19. The Case of the Telltale Fingerprint / 128
20. The Incredible Egg / 135

Psychology and Beliefs
21. Horoscope Wisdom / 142
22. Color Matters / 147
23. Memory Mystery / 152
24. Snack Foods: What Controls Your Taste? / 160

VOCABULARY AND CONCEPT EXPLANATIONS / 165

Your Senses
1. Mirror Mirror / 166
2. Flavor Detective / 167
3. What You See Is What You Get . . . Or Is It? / 168
4. Depth Perception / 169
5. Where Did I Put My Toes? / 170

Food Science
6. Food Preservation I: Dried Apples and Fruit Roll-Ups / 174
7. Food Preservation II: Refrigerator Pickles / 175
8. Enzyme Excitement in Garlic / 176
9. Testing Foods for Fat and Starch / 178
10. Snap! Crackle! Popcorn! / 179

Create Your Own
11. Boogie Woogie Butter / 180
12. Yogurt, Please / 182
13. Sprout Jungle / 183
14. Herb Garden / 184
15. Kitchen Compost / 185
16. Balance for Small Weights / 186

Inspections and Dissections
17. Water in Ketchup? / 187
18. Seed Survival / 188
19. The Case of the Telltale Fingerprint / 189
20. The Incredible Egg / 190

Psychology and Beliefs
21. Horoscope Wisdom / 192
22. Color Matters / 194
23. Memory Mystery / 195
24. Snack Foods: What Controls Your Taste? / 197

INDEXES AND APPENDIX / 199
Life Science Vocabulary and Concept Index / 199
National Science Education Standards Matrix / 200
Graphing Appendix / 204
Key to Metric Abbreviations / 210
Unmuddling Mass and Weight / 210

What's So Special About This Book?

Quite a few things! It has:

- Great activities using household supplies
- *Serious* science with a fun approach
- Clear, complete directions
- Helpful record-keeping pages
- Instruction on designing experiments
- Lots of experiment ideas, some with full directions, others where you follow your own plan (These are real experiments with plenty of data that could be used as science fair projects.)
- Vocabulary, explanations, and interesting connections at the back of the book for each activity
- Goofy jokes and inspiring quotes
- Topics, among others, in food chemistry, crime-busting, plant growth, the senses, and human psychology
- Tests of horoscopes, vision, balance, nutrition, and learning styles
- Directions for making tasty foods such as pickles, butter, yogurt, and fruit roll-ups
- Wonderful illustrations that diagram each step
- Graphing guidelines, a concept index, and information on national science standards

How to Use This Book

DOING is what the book is all about. It is the most important step toward understanding and the most fun. Look through the contents and the chapter introductions to find activities that grab your interest. Use the materials lists to see what you have at home and what you need to buy. Talk to your parents about getting the things on your list and agree on where and when you will do the activities. As you work through an activity, think about why things happen the way they do. Keep good records on the spot, while all is fresh. And, of course, clean up well at the end!

Most activities in this book have four parts. Instructions for doing the activity and space for record-keeping come at the beginning. Definitions of key terms ("Vocabulary") and explanations ("What's Going On?") are found at the back of the book. If a certain record sheet does not have enough space for you, just make your own, using the sheet in the book as a model. You may notice that parts of the "Vocabulary" and "What's Going On?" sections are found in more than one chapter. That's because certain terms and concepts come up more than once. You can skip around from activity to activity without worrying about missing a concept needed later. Each chapter stands alone.

It will be helpful to understand the following terms:

Observe (make an observation): Use your senses. Record what you feel, hear, see, taste, or smell. Example: "The corner of the cracker turned dark purple."

Hypothesize (make a tentative explanation): Try to explain how something works based on what you already know. Example: "The iodine reacted chemically with starch in the cracker."

Predict (make a prediction): Guess what is going to happen based on what you already know. Example: "The cookie is going to turn purple." Predictions and hypotheses are closely related. You make a *prediction* based on your *hypothesis*.

As you investigate, make notes and drawings of what you observe and write down your questions. A careful observer is more likely to find creative solutions to problems. Feed your brain with questions and see what you can discover. Be sure to admit to yourself when you don't "get" something. When that happens, tell yourself, "This does not make sense to me *yet*," which is much better than, "This makes no sense!" It keeps you open to learning something new. Of course, learning new things always leads to more questions. It's a lifelong process.

Life science often requires patience. Some of the activities have wait periods while organisms soak up water, grow, dry, or decay. Other activities move right along without a wait. An estimated activity time is listed to help you plan ahead. Your own time may vary, of course. Activity time is not given for extension ideas. If you are a curious and inventive person, some investigations may keep you absorbed for hours and hours.

Surprise is part of the fun of a great activity. So the instructions will not tell what will happen. Try to figure out the mysteries yourself; then when you are ready, check the "Vocabulary" and "What's Going On?" sections in the back for explanations and connections. Should you be tempted to skip these sections, think twice. You may enjoy an awesome activity, but unless you understand the concepts behind it, you haven't really gotten very far. Reading helps fill in the gaps of your experience.

If you are looking for a **science fair** activity, there are many great possibilities in this book. Just look for the symbol.

Do read "Thinking Like a Scientist" on page 7. Science, often thought of as a field of knowledge, is also a method of collecting information. Developing the skill of thinking like a scientist will enhance your life. It need not replace your ability to think like an artist or a poet or an athlete. Every skill and outlook you learn makes you more capable and more interesting.

If you want to learn more about a certain science concept, the "Life Science Vocabulary and Concept Index" on page 199 will help you find the right activities. The "Graphing Appendix" on page 204 will help you create useful graphs. And how each activity relates to the National Science Education Standards is shown in the "National Science Education Standards Matrix" on page 200. Mass and weight are compared on page 210, and the metric abbreviations used in this book are spelled out on page 210.

Enjoy your adventures with these pages. Take some risks by trying variations. And share some of your discoveries, ideas, and/or jokes with the author through her website: www.mswiz.net

A Note to Parents

May this book guide your child through hours of delight in exploring the natural world. If you are lucky enough to be invited to join in on some of the activities, model curiosity and a sense of wonder. Answer questions, but, to preserve investigative zeal, don't be too helpful, or too full of information. Reinforce your child's abilities and pose "Why?" and "What if?" questions. Enjoy the time you spend doing investigations with your child. It will promote love of learning and positive self-esteem for both of you!

Your help will be needed in collecting household, grocery, and pharmacy supplies. See page 8 for a master list of materials. Help your child brainstorm reasonable substitutions when necessary.

Notice that there is a great deal of valuable science content at the back of the book. Ideally, your youngster will gobble up new vocabulary and concepts. Be encouraging, but, if the interest isn't there just now, don't worry. The concrete experience of simply *doing* the activities builds a valuable foundation for future abstractions.

Some of the activities in this book have safety warnings, noted with the symbol:
Please read the directions carefully for all of these activities and provide supervision.

Activities that could become a **science fair** project are noted with this symbol:

Consider using activities in the book for **gatherings** of family, neighbors, scouts, and others. Activities such as "The Case of the Telltale Fingerprint," "Enzyme Excitement in Garlic," "Testing Foods for Fat and Starch," "Mirror Mirror," "Boogie Woogie Butter," and "Flavor Detective" are all good for groups to carry out. The psychology experiments "Horoscope Wisdom," "Color Matters," "Memory Mystery," and "Snack Foods: What Controls Your Taste?" also work well for group events. They require a mastermind or two to set up the experiment. The rest of the participants act as subjects.

This book is excellent for **homeschooling**. The activities and variations could be used to launch units of study. "Herb Garden" and "Kitchen Compost," for example, might begin a study of food chains or life cycles. "The Incredible Egg" and "Seed Survival" could lead into a study of reproduction. The concrete experiences create "hooks" on which to hang new information. Vocabulary and concepts are then learned more easily and remembered longer.

If you want to locate more on a certain science concept, the "Life Science Vocabulary and Concept Index" on page 199 will help you find the right activities. The "Graphing Appendix" on page 204 will help with graphs. And how each activity relates to the National Science Education Standards is shown in the "National Science Education Standards Matrix" on page 200. Mass and weight are compared on page 210, and the metric abbreviations used in this book are spelled out on page 210.

Please share your discoveries, anecdotes, ideas, and jokes with the author through her website: www.mswiz.net

A Note to Teachers

There are many ways this book can be useful for you. It can:

- be a valuable source of *classroom activities*
- provide your students with a wealth of *science fair ideas*
- form the curriculum of a *science elective*
- provide plans for after-school *science "parties"*
- be a student sourcebook for a *home science program* within your curriculum

Science Elective

If your school has electives in the schedule, you might like to develop a "Science Workshop" elective. Such a course can be as much fun for you as for the students. Choose hands-on activities that you love, and after the students have spent several days on a project, take time to process the vocabulary and concepts that have come up. Then move on to a new project. Students in an elective can do activities that are too messy or too time-consuming to use with every science student. The course can also be your testing ground for new ideas.

Science Parties

Many of the activities in this book lend themselves to optional 60- to 90-minute after-school sessions for students who sign up in advance. Students of all ability levels are enthusiastic about such fun sessions. "Boogie Woogie Butter," "The Case of the Telltale Fingerprint," "Mirror Mirror," "Flavor Detective," "What You See Is What You Get . . . Or Is It?," "Yogurt, Please," "Where Did I Put My Toes?," "Balance for Small Weights," and "Water In Ketchup?" would be particularly good for this. Take some action photos of the students loving science during your parties and send them to the local newspaper.

Home Science

You can tie home science activities into your required curriculum. Most students appreciate having options about homework and enjoy assignments that are lab oriented. Assign activity choices one night a week or assign a particular activity when it relates to class curriculum. There is never enough lab time at school and taking advantage of home time will truly enrich your course. Require a brief record of time spent, activity done, and concepts learned. You may also want to have parents initial the record. For best parental support,

send a letter home at the beginning of the program to explain the process. Assign credit based on time spent and choices made. For more information on a home science program called MOS, for "My Own Science," see the author's web site at: www.mswiz.net

Since this book was written for both school and home use, the activities are written directly to the youngster. Measuring tools and other materials listed are those most likely to be available in the home. Both English and metric units are given, so readers may choose the system they prefer.

Note that explanations are in the back of the book where they won't spoil the inquiry nature of the activities. Students can make their own unbiased observations. If you use these activities in class, you can decide when to provide the concept explanations.

Teacher tips are indicated with this icon:

To identify which National Science Education Standards are addressed by each activity, see the "National Science Education Standards Matrix" on page 200. To find activities that deal with specific topics, see the "Life Science Vocabulary and Concept Index" on page 199. For an explanation of the distinction between mass and weight, see page 210. Also note the "Graphing Appendix," page 204, and the "Key to Metric Abbreviations" on page 210.

Finally, enjoy using these materials with your students. Please share your feedback, anecdotes, ideas, and jokes with the author through her website: www.mswiz.net

Thinking Like a Scientist

Science is more than a bunch of information; it is also a method. It is a good way of learning how things work in the world and universe. Figuring out how to control high blood pressure, improve tire traction, understand the life of a star, and make computers faster are all done by experimenting. To be sure that the information collected by experiments is reliable, certain rules are followed. Even though scientists are each unique human beings, they have one trait in common: they all share great respect for the importance of these rules. The rules become a way of thinking.

The science community has never officially agreed on one list of these rules, but if it had, the list would go something like this:

The Rules
Human perceptions and memory are often mistaken. So,

1. Find ways to measure time, distance, temperature, and so on with good tools.
2. Write down measurements and other observations right away.
3. Keep an open mind until lots of data are collected.

Humans have imperfect judgment. Even when we try to be honest and fair, we are often biased. So,

4. Clearly state the question to be answered.
5. Design experiments so that they keep possible bias from affecting the results.
6. Design experiments so that only one factor is investigated at a time.
7. In the experiments, carefully keep each factor, aside from the tested factor, the same for every group of trials.

Sources of error will still creep in. So,

8. Do many trials with the same conditions so that imperfections will tend to affect all groups of trials about the same.
9. Be sure others repeat your results before you trust them too much. Maybe there was an error you could not see.
10. Be open to the idea that new information may, in the future, show truths that are not clear now.

These rules do not come naturally to most people. We don't want to write information down. "I'll remember! I saw it happen with my own eyes." We don't want to repeat what seems clear with one try. "Once is enough!" We don't want to keep an open mind. "I *know* she's lying!" So it takes practice to learn to apply the rules. This book will help provide that practice.

There are other parts to thinking like a scientist that involve traits you were born with. They are being curious and having a great imagination. Good scientists are delighted by the workings of the natural world and explore it with eagerness.

If you are curious about crime-busting with fingerprints, how much abuse seeds can handle, or how well horoscopes work, you have what it takes to enjoy this book and to practice *thinking like a scientist*.

Master List of Materials

- Apples, large, 2–3
- Ascorbic acid (Vitamin C) or lemon juice
- Avocados, 2
- Baking soda
- Bandanna
- Beans such as fava or lima, fresh, canned, or frozen
- Blender
- Bowls
- Boxes with lids, 2
- Bread or crackers
- Buckets, plastic, two 5-gallon (18-L)
- Butter
- Candy thermometer
- Cardboard, such as from the back of a pad of paper
- Celery seed
- Cheesecloth
- Chlorine bleach
- Clear tape, wide
- Coffee mugs
- Cookie sheets, 2–3, one a nonstick type
- Cooking pan
- Cooler
- Copy machine or computer and printer
- Corn syrup
- Cucumbers
- Cutting board
- Deck of cards
- Dill seed
- Dill weed or fresh dill
- Drill and 1/2-inch (1.3-cm) drill bit
- Drinking glasses
- Eggs, 10–12
- Electric mixer
- Eye dropper
- File cards, large and small
- Flashlights, 2, bright and dim
- Flower pots
- Foil
- Foil baking cups
- Food coloring
- Garlic
- Garlic press
- Glue
- Hand lens, optional
- Herb seeds
- Horoscope from a daily newspaper
- Hot pads
- Iodine tincture
- Jelly beans, 5 flavors
- Ketchup, new, in squeeze bottles
- Knife, sharp
- Lemon juice or ascorbic acid
- Masking tape
- Measuring cup
- Measuring spoons
- Microwave oven
- Milk
- Mirror, 10 inches by 10 inches (25 cm by 25 cm) or larger
- Mung bean seeds
- Music with a zippy beat
- Napkins
- Needle
- Oil, cooking
- Oven
- Paper cups
- Paper toweling
- Paper, black
- Paper, white copy
- Peaches, 29-oz. (824-g) can
- Peanuts in the shell, a few
- Pencils
- Pepper
- Permanent marker
- Plain yogurt
- Plastic bags
- Plastic drinking straw
- Plastic wrap
- Plates
- Popcorn
- Popcorn popper, optional
- Potting soil
- Quart size jars with screw-on lids, 2
- Rubber bands
- Ruler
- Salt
- Scissors
- Shovel
- Snack foods, name brand and generic
- Soft drink, clear
- Soil, from outdoors
- Soil, potting
- Spoons
- Spray bottle
- Stopwatch or timer
- Straight pin
- Sugar
- Sunny window
- Thread
- Toothpicks
- Trowel
- Whipping cream
- White vinegar

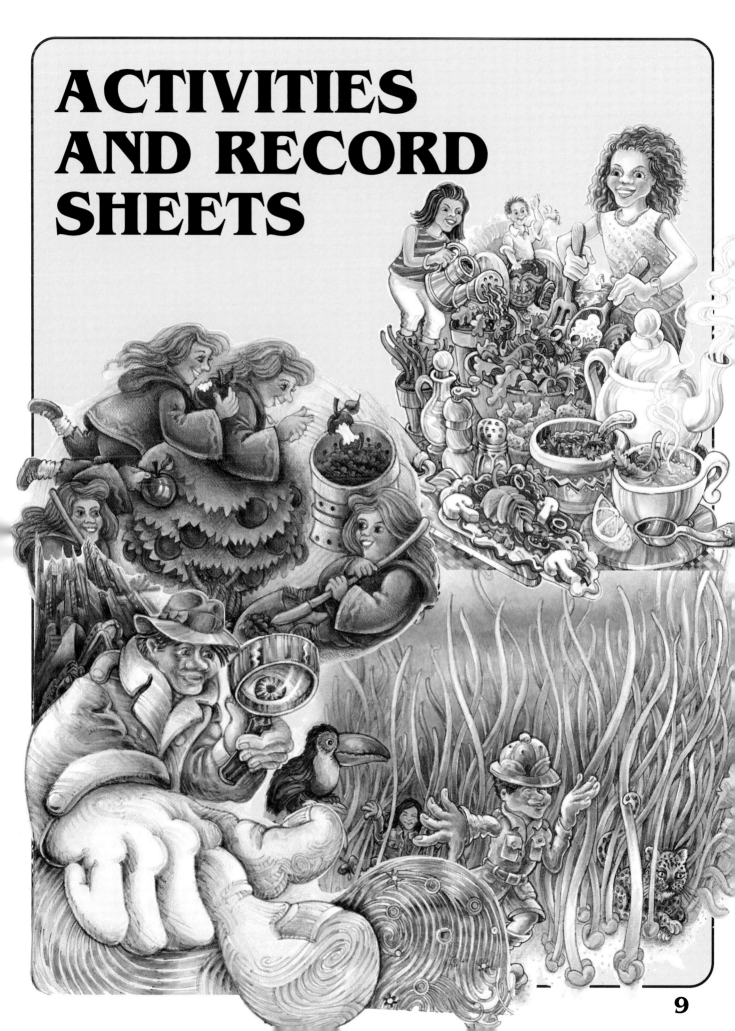

ACTIVITIES AND RECORD SHEETS

1. Mirror Mirror

Do you think you are coordinated? Try this activity and you will wonder. No, actually you will feel like a complete idiot. Yet it is definitely fun. Try it for insight into how your brain and hands work together. Watch your friends do it for laughs.

Materials

A mirror, about 10 inches by 10 inches (25 cm by 25 cm) or bigger
A weighted box or similar object to support the mirror in an upright position (A bathroom or dresser mirror may already be in a good position for you)
Tape
1 table
1 cereal box or large, thick book
Paper and pencil

I. Mirror Writing Introduction

Activity time: 10 minutes

1. Set up your writing station as shown with blank paper on the writing "stage." Make sure the mirror is well supported and/or taped into position so that it can't fall over.
2. Sit so that you can see the paper and your hand only in the mirror. The cereal box blocks your direct view of the writing stage.
3. *Close your eyes* and write your name on the paper the normal way. Not hard, right?
4. Now write your name again the normal way, but do it while watching your writing in the mirror.
5. Repeat this a few times, both with eyes open and closed.
6. Describe your experience on the record sheet.

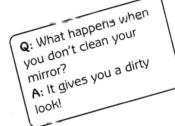

Q: What happens when you don't clean your mirror?
A: It gives you a dirty look!

II. Mirror Tracing

Activity time: 15–45 minutes

1. Place this book so that one of the "Trace These" pages (pages 13 and 14) is on the writing stage with a sheet of fresh paper over it.
2. Trace the drawings one at a time. Take your time and keep your sense of humor. As you work, think about what is going on with your coordination. What gives you trouble? What methods help you? Record your observations and thoughts on the record sheet.

> "Anyone who has never made a mistake has never tried anything new."
> —Albert Einstein

III. Mirror Writing Continued

Activity time: 10–20 minutes

1. Move this book out of the way and put a blank sheet of paper in its place.
2. Looking again only in the mirror, write your name so that it appears correctly in the mirror.
3. Do the same with other words of your choice.
4. Describe your experience on the record sheet.

IV. Investigation Ideas

Activity time: will vary

- Create your own designs to trace.
- Try writing with the base of the mirror turned 90 degrees so that it is to the *left* of your hand instead of in front of you. If you are left handed, turn the mirror so that it is to the *right* of your hand. Omit the cereal box for this and just keep your gaze on the mirror image.

Cory: Do you look in the mirror after you wash your face?
Jory: No, I look in the towel!

- See what happens if you practice mirror writing 10 to 15 minutes every day for a week or two. Date each page so that you can track your progress. You might want to record the time it takes to do a page as well.
- How does age affect mirror writing? Are 5- and 6-year-olds, new writers, better or worse at mirror tracing than other age groups?
- Do either left- or right-handed people have an advantage?
- What about doing other tasks while looking in the mirror, such as threading a needle, building with Legos, or shaping a clay figure?
- Do this as if you are right-handed, whether you are or not: Without a mirror, take a pencil in each hand and position the tips close together on a piece of paper. Write your name with both hands going in opposite directions at the same time, the right hand moving to the right and the left hand moving to the left. Each word is a mirror image of the other. What you wrote with your left hand is how Leonardo da Vinci wrote in his notebooks. Now hold what you wrote up to a mirror.

See "Where Did I Put My Toes?" on page 35 for more on these topics.

Vocabulary and concepts appear on page 166.

Trace These

14

Your Senses / Mirror Mirror

Trace These

Mirror Mirror / Record Sheet

Observations, Ideas, Sketches, Questions

Describe your reactions to the mirror-tracing and writing activities right after or in the middle of doing them. How did your brain deal with the challenge? What made it hard? What strategies were most successful?

Mirror Writing Introduction:
___

Mirror Tracing:
___

Mirror Writing Continued:
___

Investigation Findings:
___

2. Flavor Detective

It seems obvious that tasting food happens in the mouth. But does the total flavor come *just* from the mouth? Check out the effect of using and not using your nose while tasting jellybeans. Grab a buddy to work with. Sorry, kid, *ya gotta eat candy* for this job.

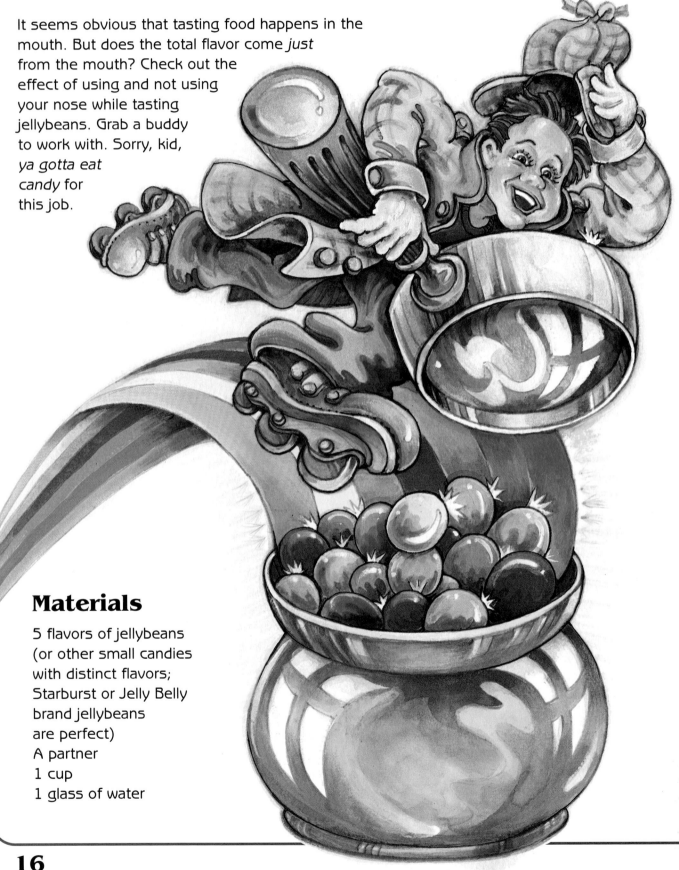

Materials

5 flavors of jellybeans (or other small candies with distinct flavors; Starburst or Jelly Belly brand jellybeans are perfect)
A partner
1 cup
1 glass of water

Flavor Detective Experiment

Activity time: about 45 minutes

The big picture: After getting familiar with the candy flavors using taste, sight, and smell, you will see how well you can identify those flavors without being able to see them, and then without being able to see *or* smell them. The hardest part is to keep from peeking or breathing through the nose while you collect the data. Just remember that being *right* when you report the flavors makes no difference in this experiment. Being *fair* in how you test them matters a lot.

1. Complete the purpose and write your prediction of the outcome on the record sheet.
2. Eat a jellybean of each color, one at a time, and agree with your partner on the name of each flavor (such as "lime," "lemon," and so on). Complete the Color/Flavor Key on the record sheet. The flavors should be clear to you at this point. If all the colors taste about the same, choose a different kind of candy.
3. Collect two jellybeans of each color in one small cup, ten in all, and mix them up. In a moment, you will be taking out beans without looking and handing them to your partner.
4. Have your partner, now called the *subject*, close his eyes. He is to keep his eyes closed until he tastes all ten jellybeans. Hand him a jellybean to taste, noting the color on the record sheet as you hand it to him. When he tells you what flavor he thinks he tastes, record that, and then put a check under "correct" or "incorrect" on the record sheet. He should take a drink of water between each flavor. DO NOT LET THE SUBJECT KNOW IF HE WAS RIGHT OR WRONG. Count only trials that are done properly. Do not change flavor decisions once they have been written down. This round is your *control group*.
5. Reverse roles, and repeat steps 3 and 4.
6. Now have your partner close his eyes again and *hold his nose firmly closed* (he holds it, not you!). He should practice breathing through the mouth and swallowing without letting air go through the nose.
7. Hand him beans as you did before, noting color and "correct" or "incorrect" WITHOUT GIVING ANY FEEDBACK. This round is the *experimental group*.
8. Reverse roles and repeat steps 6 and 7.

Jake: My dog has no nose.
Blake: Gosh, how does he smell?
Jake: Horrible!

9. Tally the *number* correct and *percent* correct for each set of conditions. To find the percent correct, divide the number correct by the number of tries and multiply the answer by 100.

   Example: If 8 out of 10 were correct, that would be 8 ÷ 10 × 100 = 80%.

10. Average the two percents correct for each set of conditions. Do this by adding the two percents together and then dividing by two. Compare the average percent correct *with* the use of the nose to the average percent correct *without* the nose.

11. Make a bar graph of the two percents, with and without the use of the nose, to show off your results. See the Graphing Appendix on page 204 for help in creating a good graph.

12. Consider any sources of error. Were all trials done fairly? Did either subject have a cold or let air through the nose? List your sources of error on the record sheet.

13. Write a one-sentence conclusion to the experiment telling the effect of smell on flavor detection.

Vocabulary and concepts appear on page 167.

Flavor Detective Experiment

Purpose:
The purpose of the experiment is to find out the effect of the sense of smell on _____.

Prediction:

_____

_____

Color/Flavor Key

	CANDY COLOR	CANDY FLAVOR
1		
2		
3		
4		
5		

SUBJECT #1   **Control Group**   Flavor Data for Eyes Closed

	CANDY COLOR	FLAVOR SUBJECT SAYS	CORRECT	INCORRECT
1				
2				
3				
4				
5				
6				
7				
8				
9				
10				

Number Correct: _____

Percent Correct: _____

Flavor Detective / Record Sheet

SUBJECT #2 Control Group   Flavor Data for Eyes Closed

	Candy Color	Flavor Subject Says	Correct	Incorrect
1				
2				
3				
4				
5				
6				
7				
8				
9				
10				

Number Correct: _____

Percent Correct: _____

SUBJECT #1 Experimental Group   Flavor Data for Eyes and Nose Closed

	Candy Color	Subject Says	Correct	Incorrect
1				
2				
3				
4				
5				
6				
7				
8				
9				
10				

Number Correct: _____

Percent Correct: _____

Flavor Detective / Record Sheet

SUBJECT #2   Experimental Group   Flavor Data for Eyes and Nose Closed

	Candy Color	Subject Says	Correct	Incorrect
1				
2				
3				
4				
5				
6				
7				
8				
9				
10				

Number Correct: _____

Percent Correct: _____

Describe what it was like to be the subject:

Calculations:
(Example: 7 correct out of 10: 7 ÷ 10 × 100 = 70%   (70% + 80%) ÷ 2 = 75%)

Average percent correct for eyes closed: _____ for eyes and nose closed: _____

Sources of Error:

_____

_____

Conclusion:

_____

_____

_____

3. What You See Is What You Get... Or Is It?

Can you imagine looking at the blood vessels in your own eyeball? Sounds gross! Did you know a part of each of your eyes is blind? You can find your blind spots and see your blood vessels without any pain, danger, or special instruments. In this activity, you will discover that the information *received* by your eye does not always match the information *perceived* by you!

Materials

A few 3 × 5 inch file cards
2 pencils
1 ruler
1 small, dim flashlight or penlight (low battery power) or 1 regular flashlight and masking tape
1 sheet of construction paper in a dark color
A dark room
1 straight pin
1 permanent marker
A friend

> **Doctor:** You need glasses.
> **Patient:** How can you tell?
> **Doctor:** I knew it as soon as you came through the window!

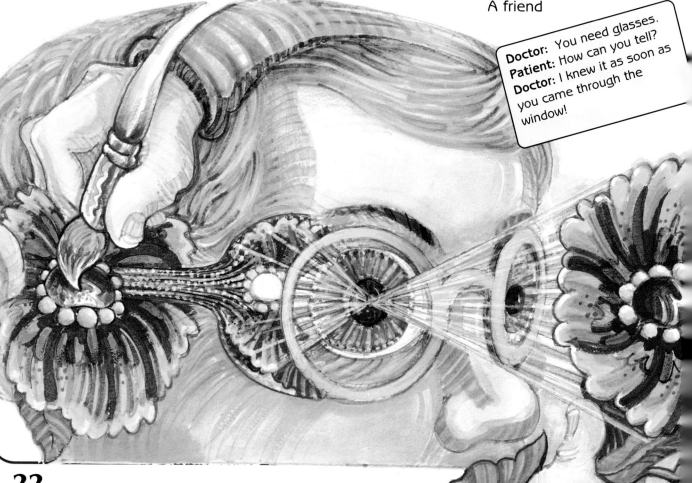

Your Senses / What You See...

I. Find Your Blind Spot

Activity time: 10–15 minutes

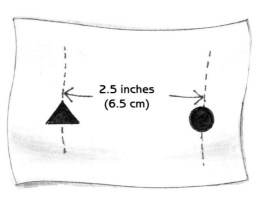

1. On a file card, draw a solid triangle and a spot about 2½ inches (6.5 cm) apart. The marks should each be about ¼ inch (0.6 cm) across.
2. Close your left eye. With your right eye, look at the triangle on the card as you hold the card out about a foot (30 cm) from your face. Notice the spot off to the right side without looking *at* it.
3. Keeping your focus on the triangle, slowly move the card closer to your face. At some point the spot will disappear. If you continue to bring the card closer it will reappear again. It may be difficult at first to see this happen so try it several times. You will "get it" eventually, and practice makes it easier.
4. Get help to use a ruler to measure the distance from the card to your eye when the spot first disappears as you bring the card closer. Record the distance along with your observations on the record sheet.

5. Rotate the card so that the spot is now on to the left of the triangle and do the same thing with the left eye. With right eye closed, hold the card out about a foot (30 cm) away. Stare at the triangle, noticing the spot as you bring the card closer to your face. Locate the position of the card when the spot disappears. Measure and record the distance.
6. Rotate the card so that the spot is directly below the triangle.
7. Repeat the process with one eye at a time.
8. Record your observations. Is it the same as when the spot was at the side?

II. Investigate Your Blind Spot

Activity time: 5–60 minutes

Think about what your observations are telling you about the eye. What questions do your observations stir up? You can make discoveries by asking a question and then figuring out how to collect data to answer it. For example,

- What would happen if you tried different shapes other than a spot?
- What would happen if you tried other card positions? (So far you tried it with the spot to the outside and with the spot below.)
- What if the two marks were twice as far apart?
- What if the two marks were half as far apart?
- What happens if you use colored paper instead of white?
- What happens if you draw a line from one edge of the card to the other through the two marks?
- How big an area can you get to disappear? Does distance from your eye matter when doing this?

Record the tests you do and the observations you make on the record sheet.

III. Find the Size of Your Blind Spot

Activity time: 15–20 minutes

You can find an approximate diameter of the blind spot inside your eyeball by making simplifying assumptions about your eye, measuring the width of the gap in your vision at a certain distance, and then using proportions (equal ratios) to calculate it.

1. Let's assume that the length of your eye from front to back is 0.92 inches (2.3 cm). Let's also assume that light travels in a straight line from pupil to retina and that the back of your eye is flat.
2. Make a marker line in the center of the left edge of a 3 × 5 card.
3. Mark the head of a straight pin with a brightly colored permanent marker and poke it into the eraser of a pencil.
4. Have someone help you with steps 5 through 7.
5. Hold the card at eye level against a wall. Use a ruler to carefully position your eyes 12 inches (30.5 cm) from the card. Your friend should help you keep this distance.
6. Stare at the mark on the left edge as you slide the pencil along the card from the right to bring the head of the pin into view. When the pin disappears into your blind area, hold the pin still so your friend can mark its location on the card.

7. Continue moving the pin (on the pencil) to the left. Make another mark where the pin comes back into view. Move the pin back and forth several times to clarify the boundaries of the blind area. Adjust the markings as needed.
8. Now use the ruler to measure the distance across the blind area between the two marks. This is its diameter.
9. The ratio of the actual blind spot diameter in your eye compared to the blind area diameter on the card is the same as the ratio of the length of your eye compared to the distance from eye to card. Write out your calculations on the record sheet.

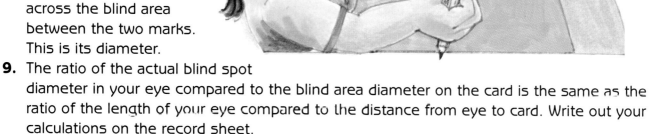

$a \div b = c \div d$

Rearrange to solve for a

$a = b \, (c \div d)$

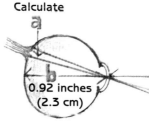

In inches: blind spot diameter = 0.92 inches (blind area diameter on card ÷ 12.0 inches)

In centimeters: blind spot diameter = 2.3 cm (blind area diameter on card ÷ 30.5 cm)

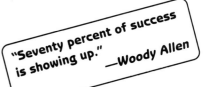

"Seventy percent of success is showing up." —Woody Allen

IV. See Blood Vessels in Your Retina

Activity time: 10–15 minutes

1. Read these directions in advance so you won't be "in the dark" when you turn out the lights.
2. If your flashlight is bright, put masking tape over parts of the front to block some of the light.
3. Find a room you can darken, such as a closet or a bathroom with no windows.

4. Turn out the lights.
5. Hold the sheet of dark construction paper out about a foot (30 cm) in front of one eye.
6. Hold the dim light up just below the eye, pointed into the eye.
7. Stare straight ahead at the paper.
8. Being careful not to poke your eye, move the light back and forth, as if searching the back of your eye. Keep staring at the paper with both eyes.

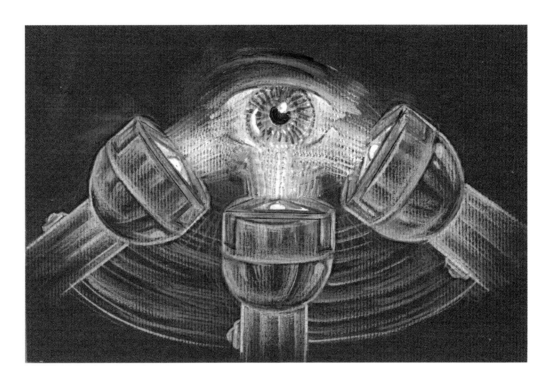

9. You are looking for branching blood vessels. They will appear on the paper like the branches of a leafless tree. It may take a little while before you notice them. Keep trying. If you don't see them, give your eyes a break for a few minutes and try again. It will work.
10. Switch to the other eye and repeat. Record your observations.

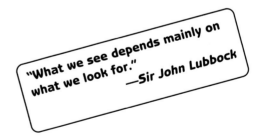

"What we see depends mainly on what we look for."
—Sir John Lubbock

Vocabulary and concepts appear on page 168.

Observations, Ideas, Sketches, Questions

ORIGINAL BLIND SPOT OBSERVATIONS

Card Position	Right Eye	Left Eye
Spot to the Outside		
Spot Below Triangle		

BLIND SPOT INVESTIGATIONS

Things I Tried	Results

Conclusions About the Blind Spot:
_____
_____
_____
_____

Blind Spot Calculations:

Spot diameter = _____ × (_____ ÷ _____)
Spot diameter = _____

Observations of Blood Vessels:
_____
_____
_____

4. Depth Perception

A friend tosses you a ball and you grab it out of the air. How did you know where to reach for it? You just *knew* and probably you take the ability for granted. But to connect with a ball in the right place, you must correctly judge distance. Without depth perception, games with moving balls would be a nightmare of missed hits, missed catches, and bops on the head. This experiment will give you some insight into how the eyes and brain work together to judge distance, or depth.

Your Senses / Depth Perception

> "Darkness cannot drive out darkness; only light can do that. Hate cannot drive out hate; only love can do that."
> —Martin Luther King, Jr.

Materials

1 cylinder-shaped container with writing on it, such as a soup can, shampoo bottle, or glue stick
1 ruler
2 identical pencils or pens
Another pencil for record keeping
A table
A partner
Paper
Tape
An 8-inch (20-cm) stack of books
A large cereal box or cookie sheet
1 red object
1 blue object

I. Look at a Can or Bottle

Activity time: about 10 minutes

1. Find a cylinder-shaped container that has writing on it such as a soup can, shampoo bottle, or glue stick.
2. Hold it upright about 6 inches (15 cm) away from your eyes.
3. Close the left eye and notice what letters or markings you can see on the far right side and the far left side of the container.

View with left eye   View with right eye

4. Without moving the can, close the right eye and again notice the limits of what you can see on the right side and the left side of the container.
5. Record your observations. Do both eyes see exactly the same scene?
6. Repeat steps 2–5 except move the can to 12 inches (30 cm) from your eyes.
7. Repeat with the can at arm's length. How do the more distant views compare to the views close up? Record your observations and complete the statements on the record sheet.

Your right eye sees this portion of can

Your left eye sees that

Near objects appear somewhat different to each eye, revealing more of one side or the other.

But far away objects appear very similar to each eye.

II. Testing Depth Perception

Activity time: 45–60 minutes

1. Your partner will be the first subject. Seat yourselves 4 to 5 feet (1.2 to 1.5 m) apart at opposite ends of a table. You need the two identical pencils, another pencil to write with, and your record sheet. Put the record sheet on a chair near you but out of the sight of your partner.
2. Create a stage for the pencils by drawing four spots on a piece of paper like this:

Tape the paper to the top of the stack of books. Put a big cereal box or a tray on its side in front of the books to hide your hands. Since "right" and "left" are different for the two of you, find something *red* to put on the side that is the *right* side for you and something *blue* to put on your *left* side. Your subject will refer to the colors when the testing begins.

3. The subject's eyes should be level with the tops of the pencils. If they are not, adjust the subject's position or the stage height as needed. You might want to have the subject rest her chin on a stack of books at just the right height to keep this variable constant. Your hands should stay out of sight behind the box, and your shoulders should stay even. Make sure there are no obvious shadows from the lighting.
4. While your partner closes or covers her eyes, position the pencils by holding them from the bottom straight up on two of the four spots. Then ask your partner to look with just the right eye. (The left may be closed or completely covered.) "Which pencil is closer?" Your partner should answer "red," "blue," or "the same."
5. Check off your partner's response on the record sheet out of her view. Don't tell anything about being right or wrong until after all the trials for that person are done. Scramble the sequence of the trials so your partner doesn't know what to expect.

Your Senses / Depth Perception

6. Repeat this process until you have completed the "Right Eye" section of the record sheet.
7. Repeat for just the left eye. Follow the list for the positions but remember to scramble their order. Keep recording your results.
8. Repeat for both eyes together.
9. Do you see a trend in the results?
10. Trade places and have your partner test you. Adjust your position as needed so that your eyes are level with the pencil tops.
11. Calculate the percent correct for the right eye alone, the left eye alone, and both eyes together.

(percent correct = number correct ÷ number possible × 100)

Example: If 4 out of 6 trials were right, 4 ÷ 6 = 0.67 and 0.67 × 100 = 67%.

12. Also calculate the average percentage correct when using one eye and when using two eyes.
13. Your data lends itself to a bar graph with two bars, one for each percentage found in step 12. See the graphing appendix on page 204 for more information on making a bar graph.
14. Write your conclusion on the effect of using two eyes on depth perception.

> Mandy: Gosh, I have this stabbing pain in my eye when I drink iced tea!
> Andy: Maybe you should take out the spoon?

III. Other Depth Perception Investigation Ideas

> "He who asks a question is a fool for a minute; he who does not remains a fool forever." —Chinese Proverb

Activity time: will vary

- Try different distances between the two pencils.
- Try different distances between the subject and the pencils.
- Try lowering the stage to table height and using something lower to cover your hands.
- Try holding one pencil in place and sliding the other slowly from far to near. The subject is to say "stop!" when the distance of the two pencils matches.
- Try a game of catch using one eye. Use a soft ball such as a tennis ball. If you can't hold just one eye open, fashion a blindfold for one eye. Compare to using two eyes.
- Try changing the catch game to rolling a ball back and forth while sitting on the floor or at a table.

Vocabulary sends and concepts appear on page 169.

Observations, Ideas, Sketches, Questions

OBSERVATIONS OF CAN OR BOTTLE

	RIGHT EYE	LEFT EYE
6 Inches		
12 inches		
Arm's length		

Complete These Sentences:

Each eye sees_____.

The closer the container is to your eyes, the _____ the view of one eye differs from the other.

DEPTH PERCEPTION TEST—PARTNER

	✓ = RESPONSE TO "WHICH IS CLOSER?" * MEANS CORRECT ANSWER									
	RIGHT EYE			LEFT EYE			BOTH EYES		
	red	blue	same	red	blue	same	red	blue	same
	*			*			*		
	*			*			*		
		*			*			*	
		*			*			*	
			*			*			*
			*			*			*
Number Correct									
Percent Correct									

Depth Perception / Record Sheet 33

DEPTH PERCEPTION TEST—SELF

✓ = RESPONSE TO "WHICH IS CLOSER?"
* MEANS CORRECT ANSWER

	RIGHT EYE			LEFT EYE			BOTH EYES		
	red	blue	same	red	blue	same	red	blue	same
	*			*			*		
	*			*			*		
		*			*			*	
		*			*			*	
			*			*			*
			*			*			*
Number Correct									
Percent Correct									

AVERAGE PERCENTAGES

	ONE EYE	TWO EYES
Total:		
Average Percentage:	(divide total by 4)	(divide total by 2)

Conclusion:
Tell the effect of using two eyes on depth perception.

_____

_____

RESULTS OF OTHER INVESTIGATIONS

What I Tried	What Happened

5. Where Did I Put My Toes?

Don't look now, but do you know where your toes are? Of course you do! And you probably never give a thought to what life would be like if you had to look around for them. Ian Waterman of Hampshire, England, knows exactly what that is like. At age 19, an infection destroyed his sense of touch and his sense of body position.* As a result, he has to look to know the location of any body part below his head. How *do* we know where our body parts are? How do we control their movements? This activity will give you insight into balance and motion control.

Materials

Yourself
A friend
A wall

*Cole, Jonathan, *Pride and a Daily Marathon*, The MIT Press, Cambridge, Massachusetts, 1995.

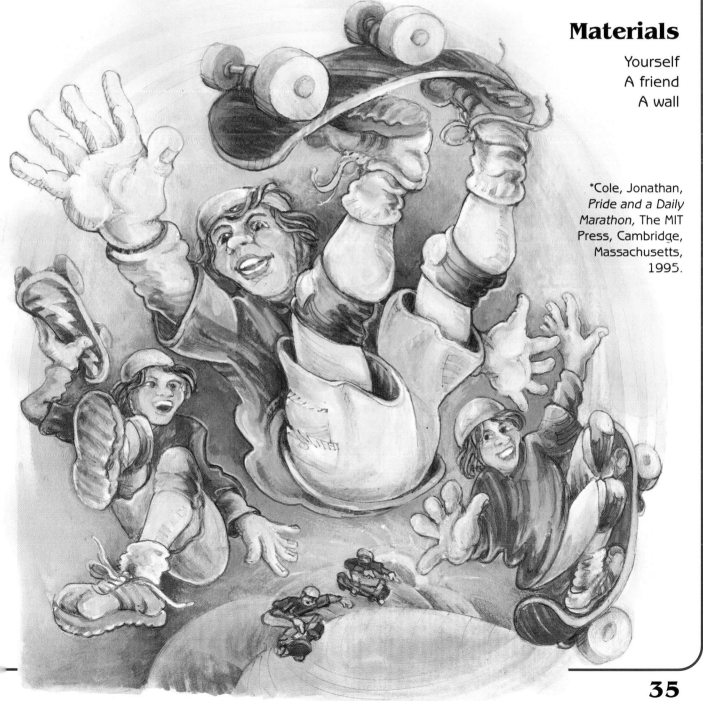

Part I. How Well Do You Know Where Your Body Parts Are?

Activity time: about 10 minutes

1. Take off your shoes and socks.
2. Close your eyes and do a mental check of your toes, feet, lower legs, upper legs, torso, shoulders, arms, hands, and fingers. Do you have a sense of where each part is? Think about how you know. What are your clues? Try to imagine what it would be like NOT to have that sense. Make some notes on the record sheet.
3. Now try to connect the following body parts with your eyes closed. In each case, aim for a specific spot of contact and see how close you are to your target. Record your results on the record sheet.
   a. Touch your two big toes together.
   b. Touch one big toe to the center of the back of the heel on your other foot.
   c. Touch the tip of your nose with your index finger.
   d. Touch your unmoving index finger with the tip of your nose.
   e. Touch your two elbows together.
   f. Touch an elbow to a kneecap.
   g. Touch the tips of your little fingers . . . in front of you, high overhead, way to one side.
   h. Touch an ear lobe with your thumb.
   i. Touch all five fingertips of one hand with the fingertips of the other hand. Compare this to connecting one pair of fingertips at a time. Try both of these centered in front of you and also way off to one side.
   j. Try other points of contact and variations.
4. What patterns have you noticed? Record your observations and insights on the record sheet.

Ted: Do you know what has a bottom at the top?
Ed: Nope.
Ted: Your legs!

Part II. Try These Moves!

(Careful! You are Entering The Unbalanced Zone)

Activity time: about 10 minutes

Try each of the following actions exactly as described and record your results on the record sheet.
1. Stand with your backside and heels against a wall. Try to bend over slowly and touch your toes; then return to standing.

Your Senses / Where Did I Put My Toes?

2. Stand against a wall so that your toes and the tip of your nose touch it. Now, keeping contact with the wall, try to stand up on your tiptoes.
3. Sit on a straight-backed chair and position your feet so that your knees and hips make right angles. Have your friend hold a hand just in front of your forehead. Try to stand up from that position without bumping into your friend's hand.

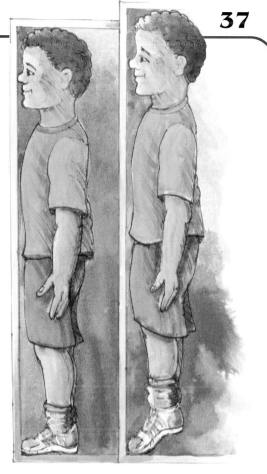

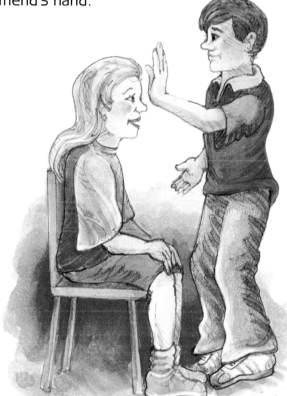

"Argue your limitations and sure enough they're yours."
—Richard Bach

4. Stand with your left side and left foot against a wall. Try to pick up your right foot.
5. Kneel on the floor and put one elbow against your knee extending your forearm out in front of you on the floor. Stand a video or small paperback book centered in front of you at the point just beyond your fingertips. Now clasp your hands behind your back and try to knock the video over with your nose.

Part III. Watch Your Partner Do the Moves

Activity time: about 5 minutes

Feeling like a loser? Don't worry. You can figure out the source of your troubles.

1. Carefully watch your friend do the following motions. Note on the record sheet any little shifts in body position beyond those described in the instructions. Watch from the side for motions *a* through *e* and from the front for *f*.
   a. Bend down to touch your toes. What happens to the rear end?
   b. Go up on your tiptoes. What is done first?
   c. Sit on a chair and position your feet so that your knees and hips are making right angles. Stand up. What must be done before standing?
   d. Lift one leg high in front of you. What happens to the back?
   e. Put on a heavy book bag. How does the body position change?
   f. Stand normally. Pick up your right foot. Watch from the front. What is done before the foot is picked up?
2. Switch places with your partner so you both get to see what these movements require.
3. Summarize on the record sheet what was similar about all these situations.

Part IV. What Happens When We Spin?

Activity time: about 10 minutes

1. If your friend is willing, find a safe area and have her spin around until she is dizzy. You can use a spinning chair if you have one. Agree in advance on an obvious spot on the wall that she will point to for 30 seconds as soon as she stops spinning. Stand behind her to judge the direction of her pointing once she stops spinning. Record your observations on the record sheet.
2. Again have your friend spin around until dizzy. This time look into her eyes when she stops. What do you observe? Record your observations on the record sheet.
3. Switch roles and spin yourself. Record your observations on the record sheet.

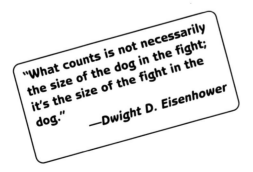

"What counts is not necessarily the size of the dog in the fight; it's the size of the fight in the dog." —Dwight D. Eisenhower

Warning!

Make sure the area is clear so no one trips or gets bumped.

Vocabulary and concepts appear on page 170.

Observations, Ideas, Sketches, Questions

Part I. Data

Notes on imagining the location of your body parts: _____

_____

_____

CHECK HOW WELL YOU HIT THE TARGET

	PERFECT	**CLOSE**	**OFF**	**WAY OFF**
Toe to toe				
Toe to heel				
Finger to nose				
Nose to finger				
Elbow to elbow				
Elbow to kneecap				
Little fingers in front				
Little fingers high				
Little fingers to the side				
Thumb to ear lobe				
5 fingertips to 5 fingertips front				
5 fingertips to 5 fingertips side				
Fingertips one at a time front				
Fingertips one at a time side				

Other things you noticed about the experience: _____

_____

Part II. Data

CHECK THE PROPER COLUMN

	Did It	Couldn't Do It
Bend over		
Up on toes		
Stand up		
Pick up foot		
Knock over video		

Part III. Data

OBSERVATIONS OF BODY ADJUSTMENTS

a. Bend over	
b. Go up on tiptoes	
c. Sit on a chair, then stand up	
d. Lift leg in front	
e. Put on book bag	
f. Pick up right foot	

Summary/Insights: _____
_____
_____

Where Did I Put My Toes? / Record Sheet

Part IV. Data

YOUR OBSERVATIONS

The spinner's pointing	
The spinner's eyes	
Your experience after spinning	

6. Food Preservation I

Dried Apples and Fruit Roll-Ups

Tammy: Why are you staring so hard at that can of apple juice?
Sammy: It says "concentrate!"

Drying is an ancient method of keeping food from spoiling. Slices of fresh apple will brown in the air and mold in a bag. But apple slices dried after soaking in a special solution stay fresh for weeks. And they make a great, light snack that can go anywhere. You get bite-size pieces with no core, no mess, and no waste. Yum! With fruit leather roll-ups you can be creative by combining fruits and adding flavorings and fillings. Yum^2 *(that's Yum Yum)*.

Food Science / Food Preservation I

Materials for Dried Apple Slices

An oven that can be set to 150°F (66°C) (Use a food dehydrator if you have one.)
2 or 3 large apples
Water
1 sharp knife

WARNING!: Be careful!

Vitamin C tablets (ascorbic acid)—six 500 mg tablets—or 2 cups (480 mL) of lemon juice
1 sturdy drinking glass
1 bowl
1 cutting board
1 spoon

Paper towels
1 nonstick cookie sheet (a bare metal sheet may blacken the apples)
1 plastic bag
1 permanent marker

> "It's not that I'm so smart, it's just that I stay with problems longer."
> —Albert Einstein

Materials for Fruit Leather Roll-Ups

An oven that can be set to 150°F (66°C) (Use a food dehydrator if you have one.)
1 blender or food processor
1 29-ounce (824-g) can of peaches, drained (or 2 cups (480 mL) of any of the following fresh, frozen, or canned fruits: apricots, nectarines, apples, cherries, pears, strawberries, raspberries. Choose a single fruit or use a combination.
2 teaspoons lemon juice or $1/8$ teaspoon (375 mg) powdered ascorbic acid
2 cookie sheets (check that they are quite flat)
Plastic wrap
Tape

I. Making Dried Apple Slices

Activity time: 30–40 minutes, plus 5–12 hours drying time

The big picture is that you will prepare apple slices, soak them in a liquid that will prevent browning, dry them in the oven, cool them, and bag them. . . . Oh, yeah, then you eat them!

1. Wash your hands with soap.
2. Make the anti-browning bath. Crush six 500 mg vitamin C tablets in a bowl using the bottom of a sturdy drinking glass. (Don't hit them, just press and twist the glass against the tablets.) Stir the powdered vitamin into 2 cups (480 mL) of water until it all dissolves. If you prefer, use 2 cups of lemon juice in place of the acid solution.

3. Using a cutting board and a sharp knife, cut the apples into quarters, peel the skin off each quarter, and cut out the core. There is a faint line where the core meets the rest of the apple. Remove the core all the way to the line.
4. Cut each section into slices about 1/4 inch (0.6 cm) thick. Put the slices directly into the anti-browning bath and let them soak for 5 minutes.
5. Remove the apples with a spoon and place them on paper toweling to absorb the excess solution.
6. Arrange the slices on the nonstick cookie sheet so that they do not touch one another. Put the tray into the oven set at 150°F (66°C). If the temperature is too high, the apples will cook instead of dry.
7. Record the time you put the slices in the oven. Drying in most ovens will take 5 to 12 hours. Drying time varies with the oven, the thickness of the slices, and the amount of food being dried. Convection ovens have a fan that speeds up evaporation and take only about half the time.
8. It is helpful to open the oven door now and then. Evaporation will be speeded up by replacing the moist air in the oven with drier room air.
9. After 2 hours, check the apple slices and turn them over. If you don't have enough time to finish the drying in one day, leave the apples inside the oven, turn the oven off, and turn it back on the next day.
10. Test the apples for dryness by removing a few pieces and cutting them in half. They are ready if there is no liquid visible when you squeeze the cut pieces. The slices should still be flexible. If the apples are over-dried, they get crisp like chips but are still tasty.
11. Remove the tray from the oven when the apples are ready.
12. Leave the apple pieces on the trays just long enough for them to cool to room temperature.
13. Put the slices into a plastic bag. Write the date, the oven time, and the oven temperature on the bag as well as on the record sheet.
14. Enjoy snacking!

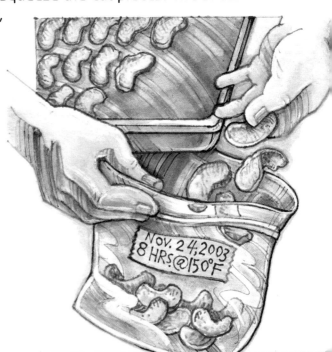

"Be ready when opportunity comes. Luck is the time when preparation and opportunity meet."
—Roy D. Chapin, Jr.

Food Science / Food Preservation I

II. Making Fruit Leather Roll-Ups

Activity time: 10–20 minutes, plus 6–15 hours drying time

The big picture is that you will liquefy the peaches or other fruit in the blender and pour the liquid fruit into a thin layer on plastic wrap to dry in the oven.

1. Lay a sheet of plastic wrap onto each cookie sheet. Smooth the plastic and tape the corners of the wrap down.
2. If you have chosen a fresh fruit, wash it, peel it, and remove any inedible seeds.
3. Put about 2 cups (480 mL) of fruit into the blender or food processor.
4. Add 2 teaspoons (10 mL) of lemon juice (or 375 mg of ascorbic acid) to prevent darkening.
5. Blend the fruit and lemon juice until smooth.
6. Taste the pureed fruit. If you have used fresh fruit and want it sweeter, add a little corn syrup or honey. Peaches canned in syrup will not need extra sweetener.
7. Pour half of the liquid fruit onto the plastic wrap of each tray and spread it into circles about 1/8 inch (0.3 cm) thick and 11 inches (28 cm) in diameter. Shake the trays back and forth a bit to level out the fruit thickness.
8. Put the sheets into the 150°F (66°C) oven. Record the time you begin the drying. Drying will take 6 to 15 hours. If you don't have enough time to finish the drying in one day you can turn off the oven, leaving the fruit wrap inside, and turn the oven back on the next day.
9. Test the fruit for readiness by pressing the center with a finger. If your finger makes a dent, dry the fruit longer. If the fruit is firm, it is time to remove it from the oven. The fruit layer should still be flexible.
10. If you like, spread or lightly sprinkle the goody of your choice on the fruit before rolling. Consider peanut butter, cream cheese, marshmallow fluff, nuts, seeds, or chocolate. Just plain might be best of all.

11. While the fruit is still warm, peel it from the plastic and roll it up. It will keep for 3 to 4 weeks in plastic wrap without refrigeration. Refrigerate any roll-ups that have cheese filling.

III. Investigations

Activity time: will vary

Other fruits and other drying techniques can be used. For example, you can:

- Make raisins from grapes. Before drying, dip whole grapes into boiling water for 30 seconds, then plunge them into ice water. Drain on toweling. This treatment speeds drying by breaking the skins.
- Dry bananas in a fashion similar to apples but set the oven at 140°F (60°C). Choose ripe, yellow bananas. Peel and cut either across or the long way into 1/8-inch (0.3-cm) slices. Soak in ascorbic acid (vitamin C) solution or lemon juice and continue as with apples but dry until crisp.
- Find the water content of any fruit. See "Water in Ketchup?" on page 114 for directions.
- Try adding flavorings to fruit roll-ups while the fruit is in the blender. Cinnamon, pumpkin pie spice, vanilla, or almond extract are possibilities.

IV. Design Your Own Experiment

Design an experiment on either apple slices or fruit roll-ups by choosing one variable, adjusting it to two levels, and comparing the outcomes. Possible variables include:

- The type of apple, such as Red Delicious vs. Macintosh.
- The way the apple is sliced, such as rings vs. wedges.
- The thickness of the apple slices.
- The use of lemon juice vs. ascorbic acid solution.
- The diameter of the roll-ups.
- The thickness of the roll-ups.
- The oven temperature. Most sources recommend 140°F (60°C). At what temperature will apples cook instead of dry?

You could use the balance from "Balance for Small Weights," page 109, to find the percent water loss per hour from two or three slices of fruit. (See "Water In Ketchup?") This data would be well suited for a line graph. See the Graphing Appendix on page 204 for help in making a line graph.

Use the outline on page 48 to write up the experiment.

Vocabulary and concepts appear on page 174.

Observations, Ideas, Sketches, Questions

Fruit	Temperature and Treatment	Drying Time	Observations

Investigation Results:

Fruit Drying Experiment

Purpose:

The purpose of this experiment is to find out _____

_____

Prediction:

_____

Procedure:

1.
2.
3.
4.
5.

Data and Observations:

Sources of Error:

_____
_____
_____

Conclusion:

_____
_____
_____
_____

7. Food Preservation II

Refrigerator Pickles

Dill pickles. Sweet pickles. Kosher pickles. Tomato pickles. Watermelon rind pickles. Pickled pig's feet! It seems there's no end to what can be pickled. The perfect place to start your pickling career is with cucumbers. Set up an experiment to find out the effects of vinegar and salt on slices of cucumber. Then try the delicious recipes here. You can make them any time of the year. Pucker up!

Q: Why did the cucumber blush?
A: Because he saw the salad dressing.

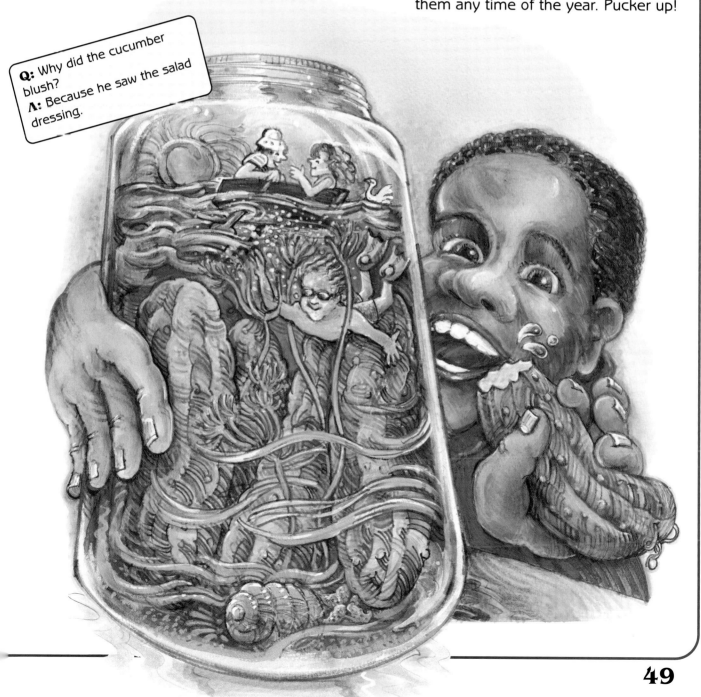

Materials for Cucumber Preservation Experiment

2 regular salad cucumbers or 4 or 5 pickling (Kirby) cucumbers (they are smaller)
Measuring spoons
1 teaspoon salt
1 tablespoon white vinegar
6 plates (3 matching pairs), ceramic or plastic, not paper
6 small plastic bags
1 permanent marker
Masking tape
A timer

Materials for Sweet Pickles

1 small onion, peeled and thinly sliced
1 pound cucumbers (4 to 6 pickling cucumbers or 1 to 3 salad cucumbers)
1 1/3 cups white vinegar
1 1/3 cups sugar
1 teaspoon salt
1 teaspoon mustard seed
1/2 teaspoon celery seed
1/2 teaspoon tumeric

Equipment for Both Pickle Recipes

Measuring cup
Measuring spoons
Quart jar with lid
Small cooking pot
Sharp knife
Cutting board

Materials for Dill Pickles

One pound of cucumbers (4 to 6 pickling cucumbers (best) or 1 to 3 salad cucumbers)
1 cup white vinegar
1 cup water
1 tablespoon salt
1 tablespoon sugar
1/4 teaspoon dill seed
1/4 teaspoon celery seed
1/4 teaspoon pepper
1/2 teaspoon dried dill weed or two sprigs of fresh dill
2 cloves (sections) of garlic

I. Cucumber Preservation Experiment

Activity time: about 30 minutes the first day plus a 2-hour wait; 5 minutes a day for a week or more after that

The big picture is that you are going to expose cucumber slices to microbes in the air and then treat some of them with salt, treat some with vinegar, and leave some plain. You'll make observations after 2 hours and then store half of each sample in the refrigerator and half at room temperature, checking them regularly for a week or so.

1. Peel the cucumbers. Cut them into round 1/4-inch (6-mm) thick slices. Throw away the slice from each end. Deal the rest out to three plates, about ten slices to a plate arranged in one layer.

2. Let the cucumber slices sit out for 10 minutes once they are all in place. During this time, microbes from the air will land on them. This was happening while you were cutting them, too, since microbes settle out of the air 24 hours a day.
3. When 10 minutes are up, leave plate #1 alone. It is your control group. Label it "Plain" with marker on masking tape.
4. On plate #2, sprinkle 1 teaspoon of salt over the slices. Label it "Salt."
5. On plate #3, dribble 1 tablespoon of white vinegar over the slices. Label it "Vinegar."

6. Cover each of the three plates with a second matching plate turned upside down, to prevent evaporation of the juices.
7. Set a timer for 120 minutes (2 hours) and let the plates sit.
8. When the timer goes off, lift off the plate covers and make observations of each sample. Carefully tilt each plate and collect any liquid that drips off into your tablespoon. Estimate the volume you collect, record your data on the record sheet, and *return* the liquid to its plate. *Wash and dry* the measuring spoon between samples so you don't transfer vinegar or salt from one plate to the next. Now you are finished with the short-term study and are ready to set up for the longer term study.

9. Use the marker to label 6 bags as follows: "Fridge Plain," "Fridge Salt," "Fridge Vinegar," Room Plain," "Room Salt," and "Room Vinegar."
10. Divide the slices from the "Plain" plate into each of the two "Plain" bags. Divide the salted slices between the two "Salt" bags and so on. For any sample that has liquid, divide that liquid evenly between its two bags. Close up the bags leaving some air space inside.

11. Stand the "Fridge" bags in the refrigerator so they can't leak.
You could put the three bags into another container to help them stand up.
12. Do the same for the other three bags, but put them in a cabinet.
13. Complete the statement of purpose on the record sheet. Think about the treatments and make some predictions about the outcomes. Record your predictions on the record sheet.
14. Without opening the bags, check all samples once a day for 3 days, then every couple of days after that. Record your observations of color and general appearance on the record sheet. Do not taste the samples. If a sample gets gross with microbe growth, record your observations of it; then throw the bag away without opening it. While many microbes are harmless, some microbes can make you sick. Continue your observations of the remaining samples.
15. Changes will continue to occur after 8 days. Decide how long you want to continue your study. Use additional paper to add to your data table as needed.
16. Consider any sources of error you may have had and record them on the record sheet.
17. What do you conclude about the effects of vinegar, salt, and refrigeration on cucumber preservation? Write your conclusion on the record sheet.

> "The pessimist sees difficulty in every opportunity. The optimist sees the opportunity in every difficulty."
> —Winston Churchill

Food Science / Food Preservation II

II. Dill Pickles

Activity time: About 20 minutes

1. Wash your hands with soap.
2. Wash the jar with hot soapy water and rinse *thoroughly*.
3. Wash the cucumbers with cool water.
4. Slice off the ends of the cucumbers. Sometimes they are bitter.
5. Slice the cucumbers into disks 3/8 inch (1 cm) thick and put them into the jar.
6. Peel the garlic cloves, cut each of them into five slices, and add them to the cucumbers.
7. Put the rest of the ingredients into the pot. Stir and heat until the liquid boils. Remove from the heat and let cool for 5 minutes.
8. Pour the hot liquid into the jar over the cucumbers and garlic.
9. Put the lid on the jar and refrigerate.
10. The pickles are ready in 2 days. But they are pretty tasty even after 1 day.

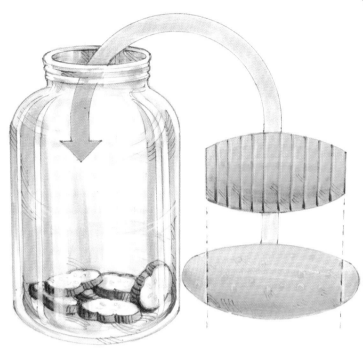

Warning: Boiling water can cause serious burns. Use caution!

III. Sweet Pickles

Activity time: about 15 minutes

1. Wash your hands with soap.
2. Wash the jar with hot soapy water and rinse *thoroughly*.
3. Wash the cucumbers well and cut off the ends.
4. Cut the cucumbers into 3/8-inch (1-cm) round slices.
5. Peel the onion and slice it thin.
6. Put the cucumber slices and onion, mixed up, into the jar.
7. Put the rest of the ingredients into the pot. Over medium heat, stir until all the sugar dissolves.
8. Pour the hot liquid over the cucumbers in the jar.
9. Close the jar and put it in the refrigerator. Your pickles are ready in 24 hours!

Warning: Use caution with hot liquids!

IV. Ideas for Further Cucumber Investigations

Activity time: will vary

- Try the recipes with other vegetables such as zucchini, green tomatoes, or partly cooked carrots.
- Grow some cucumbers outside from packaged seeds following the directions on the pack. They are easy to grow. Or take seeds from a large, fresh cucumber, dry them, and plant them outdoors or in a pot of soil indoors.
- Find an adult who can teach you how to "can" foods and make pickles that do not need refrigeration before opening.
- Test the preserving power of other food substances. Does salt substitute work? Oil? Lemon juice? Sugar? Baking soda?

"You cannot do a kindness too soon, for you never know how soon it will be too late."
—Ralph Waldo Emerson

Vocabulary and concepts appear on page 175.

Cucumber Preservation Experiment

Purpose:

The purpose of this experiment is to find the effects of _____, _____, and _____

on _____.

Predictions:

_____

_____

Data Table:

	PLAIN, ROOM	SALT, ROOM	VINEGAR, ROOM
2 Hours			
24 Hours			
2 Days			
3 Days			
5 Days			
8 Days			

Food Preservation II / Record Sheet

	PLAIN, FRIDGE	SALT, FRIDGE	VINEGAR, FRIDGE
2 Hours			
24 Hours			
2 Days			
3 Days			
5 Days			
8 Days			

Sources of Error:
_____
_____
_____

Conclusion:
_____
_____
_____

8. Enzyme Excitement in Garlic

Here's a mystery for you: A chef prepares two chicken recipes identical in every way except that one has 2 cloves of garlic and the other has 40 cloves of the same kind of garlic. Yet the 2-clove chicken has more garlic flavor than the other! How can this be? Try this experiment to find the surprising solution.

Materials:

- 3 cloves (sections) of garlic of the same size
- 3 small plates
- A garlic press (or a sharp knife)
- 1 stick of butter or margarine, cut into three equal sections and warmed to room temperature
- Masking tape
- Pen
- Bread or crackers
- Microwave oven
- A sniffing and tasting panel: you plus as many friends and family as you can recruit

WARNING: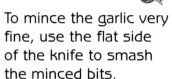

To mince the garlic very fine, use the flat side of the knife to smash the minced bits.

Q: Why couldn't Mozart find his friend?
A: Because he was Haydn.

I. Enzyme Excitement in Garlic Experiment

Activity time: about 1 hour

The big picture is that you are going to prepare three samples of garlic butter, each made with one garlic clove treated as follows. The first sample of garlic is crushed but not cooked, the second garlic sample is cooked before being crushed, and the third sample is cooked *after* being crushed. Your taste panel will describe the odor and taste and rate the intensities of odor and taste at several times during the preparation of the samples.

1. Wash your hands with soap.
2. Prepare three masking tape labels and put them on the underside of the plates: "1. Crushed Only," "2. Cooked, Then Crushed," and "3. Crushed, Then Cooked."
3. Make three smaller labels: "1," "2," and "3" to put on the upper edge of the same plates where they can be seen.
4. When several taste tests are done in a row, the first item tasted could get rated differently than other items just because it is first. The way to avoid such an *order effect* is to make sure that each item is tasted first just as often as the others. To do this, make mini record sheets like the following for each member of your tasting panel. Label the top three columns differently (1, 2, 3 or 2, 3, 1 or 3, 1, 2) for each so that each person tests the samples in a different order. Give a sheet to each of your tasting panel friends. The little boxes are for strength ratings on a scale of 1 to 10.

DESCRIPTIONS AND STRENGTH RATINGS

Sample number						
Whole clove odor						
Smashed clove odor						
Butter odor						
Butter flavor						

Food Science / Enzyme Excitement in Garlic

5. Choose three cloves of garlic that are about the same size. Peel off the dry "skins," without scratching the underlying cloves, and put one on each plate. Move the plates to different parts of the room or house so the smell of one won't affect the smell of another.
6. Testing in the assigned order, you and the rest of your taste panel should sniff the garlic samples and describe their odors on the first row of the mini record sheet. Also give each sample an odor strength rating from 0 to 10 with 0 being no odor and 10 being the strongest odor of garlic you could imagine. Each person should decide on his or her rating without knowing the ratings of the others. You get more reliable results the more subjects you have.
7. Put the first clove through the garlic press. Put the crushed garlic back on its plate.
8. Wash the press and your hands.
9. Put the second clove into the microwave for 30 seconds. Then put it through the garlic press and put the crushed garlic back on its plate. Wash the press and your hands.
10. Crush the third clove of garlic in the garlic press. Put the crushed garlic back on the plate and microwave it for 30 seconds. Wash the press and your hands.
11. Move each of the three samples to their different remote locations again so you can smell them individually. Your friends should not know how each sample was treated. You and your taste panel are going to sniff and then describe and rate the odors from each plate on the record sheet as before. But first, the panel needs to go outside and run around for 5 minutes. You can better trust your nose when it is refreshed! Panel members should complete the second row of their mini record sheets.
12. Mix one portion of butter into each of the garlic samples on the labeled plates. Be sure that the garlic is mixed well with the butter. Butter some bread or cracker with each sample and put the pieces on the plate.
13. Now return the samples to their remote locations again. Have each subject continue to test the samples in the assigned orders. Do odor and taste descriptions and strength ratings for each sample and record on the third and fourth rows of the mini record sheets.
14. Transfer the information from each mini record sheet onto the main record sheet. Find the average odor and flavor ratings at each point by adding the ratings and dividing by the number of subjects.
15. Consider any sources of error in your study; anything big or little that made the test unfair. Were the sizes of the cloves identical? Did you wash the garlic press between each sample? Could the members of your panel have been influenced by each other?

16. Make bar graphs to show your results. In one graph use three bars to show the average garlic butter flavor ratings for each of the three garlic treatments. In another, use bars to compare the average odor ratings before and after treating each clove of garlic. This will give you six bars. For help in making a bar graph, go to the Graphing Appendix on page 204.
17. What is your conclusion about the effect of no cooking vs. cooking before crushing vs. cooking after crushing? Write it on the record sheet.
18. In this study, you found out *what* happened, but you have no data on *why* it happened. Come up with a reasonable hypothesis to explain why the results came out as they did.
19. You should now be able to solve the mystery of the strong two-clove of garlic chicken vs. the mild 40-clove of garlic chicken dish. What is your solution? Write it on the record sheet.

For an easy 40-clove of garlic chicken recipe, see page 177.

"A problem well stated is a problem half solved."
—Charles F. Kettering

II. Ideas for Further Investigations

Read the explanation on page 176 to understand how the different flavor strengths of garlic come about. It will help you in designing further experiments.

- Prepare two versions of a recipe that calls for garlic, such as the chicken recipe on page 177 or spaghetti sauce. Use many whole cloves of garlic in one version and one minced clove of garlic in the other. Have your family do a taste test.
- Find out what *temperature* it takes to destroy the flavor trigger. We know that a room temperature of 70°F (21°C) doesn't hurt it. What about 100°F (38°C), 125°F (52°C), 150°F (66°C), 175°F (79°C), and so on? You could set this up with coffee cups of water heated to different temperatures. You would need a thermometer to record water temperatures and you would need to be very careful working with hot liquids. Soak the garlic cloves in the heated water for 2 minutes; then remove them with a spoon, mince them and test them. Include garlic soaked in room-temperature water for comparison.
- Similar studies could be done on the effects of other treatments. Does freezing garlic destroy the active chemical? What about soaking the garlic in salt water or in vinegar?
- You could focus on the senses of smell and taste. If tasters pinch their noses shut, how do they describe the taste of garlic-butter? (See "Flavor Detective" on page 16.)
- How long does it take for your nose to quit detecting a smell? This is called *odor fatigue*. Have a sample of an odor source in front of a friend who has eyes closed. Note the time when, breathing normally, a subject begins smelling it and when he believes you have taken the source away. Does the time it takes for odor fatigue to occur depend on the source? You could compare garlic with onions, smashed ripe banana, cinnamon, vanilla, mouthwash, perfume, and so on.

"The more I study, the more I know.
The more I know, the more I forget.
The more I forget, the less I know.
So why study?"
—author unknown

Vocabulary and concepts appear on page 176.

…
Enzyme Excitement in Garlic Experiment

Purpose:
The purpose of this experiment is to see how crushing and cooking affect the flavor and aroma of garlic.

OBSERVATIONS AND RATINGS

		Clove 1.	Clove 2.	Clove 3.
Garlic odor *before* treatment and odor strength rating	Subject 1			
	Subject 2			
	Subject 3			
	Subject 4			
	Average Rating			

		Clove 1. Crushed Only	Clove 2. Cooked, Then Crushed	Clove 3. Crushed, Then Cooked
Crushed garlic odor *after* treatment and odor strength rating	Subject 1			
	Subject 2			
	Subject 3			
	Subject 4			
	Average Rating			

Enzyme Excitement in Garlic / Record Sheet

		Clove 1. Crushed Only		Clove 2. Cooked, Then Crushed		Clove 3. Crushed, Then Cooked	
Odor of garlic in butter with odor strength rating	Subject 1						
	Subject 2						
	Subject 3						
	Subject 4						
	Average Rating						

		Clove 1. Crushed Only		Clove 2. Cooked, Then Crushed		Clove 3. Crushed, Then Cooked	
Flavor of garlic in butter with flavor strength rating	Subject 1						
	Subject 2						
	Subject 3						
	Subject 4						
	Average Rating						

Enzyme Excitement in Garlic / Record Sheet

Sources of Error:

_____
_____
_____

Conclusion:

_____
_____
_____

Hypothesis on Why:

_____
_____
_____

Solution to the Garlic Chicken Mystery:

_____
_____
_____
_____

9. Testing Foods for Fat and Starch

Researchers study foods with high-tech tools to find out exactly what nutrients, vitamins, and minerals they contain. But fat and starch in foods can be detected easily with tests you can do in a flash. Predict the contents of your favorite foods before you test them. This activity will tune you in to the chemistry of foods that you eat every day.

Food Science / Testing Foods for Fat and Starch

Materials

2 sheets of copy paper
Aluminum foil
Tincture of iodine, also called iodine tincture (available at pharmacies in the first-aid section)

1 eye dropper
A variety of food samples
1 pen
1 sturdy drinking glass

Warning: Iodine is toxic. Keep it out of reach of animals and toddlers.

> "In matters of style, swim with the current; in matters of principle, stand like a rock."
> —Thomas Jefferson

I. Getting Ready

Activity time: 10–15 minutes

1. Raid the refrigerator and pantry to collect 15 or more foods to test. You might try bread, cream cheese, jam, cereal, eggs, milk, pasta, orange juice, salad dressing, yogurt, peanut butter, olives, tomatoes, chocolate, baloney, crackers, bananas, apples, cantaloupe, nuts, and seeds, among others.
2. Before doing any tests, write down the names of the foods on the table on the record sheet and predict which ones will contain fat or starch. Show your predictions by circling one, both, or neither of "Fat" and "Starch."

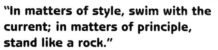

Q: Why are chefs cruel?
A: Because they beat the eggs, whip the potatoes, and cream the corn.

II. Testing for Fat

Activity time: about 10 minutes

The big picture is that you will rub bits of the foods on labeled sections of absorbent paper. Once any water in the food has had time to dry, wet-looking, see-through spots on the paper tell you which foods contain fats or oils.

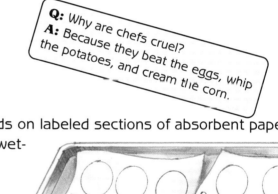

1. Draw as many circles as you have foods on one or two sheets of paper. Below each circle, write the name of one of the foods so that there is a place for each food you have chosen. Put the same names on the table of the record sheet.
2. Put the paper or papers on a tray so that they will be easy to move later without upsetting the foods you applied.
3. Drip, rub, spread, and/or crush a small sample of the foods (room temperature or warmer) in the proper circles on the paper. A few drops of a liquid or a peanut-sized solid is enough. Dry foods such as nuts won't give an accurate test unless they are crushed. Press down firmly with the bottom of a small glass to crush them into the paper. Use a clean glass for each item you must crush. Leave a piece of any solid food sitting in the circle.

4. Continue until you have applied all the foods.
5. Allow the papers to dry for an hour in a warm spot. A longer wait, even overnight won't hurt. If much fat is present, the paper will turn and stay translucent (partly see-through) around the food sample.
6. Record your results on the record sheet. A plus sign (+) means that fat was detected and is called a positive test. A negative sign (–) means no fat was detected. If you are not sure how to rate a test result, record a question mark.
7. How did your predictions compare with your test results?

III. Testing for Starch

Activity time: about 10 minutes

The starch test is an instant chemical test. Add iodine to a food sample and see if a color change to purple or blue/black occurs.
1. Work on foil for this set of tests. Draw and label circles on the foil, one circle for each food.
2. Place a sample of the proper food in each circle. It need not be crushed or smeared.
3. Use the eye dropper to take up some of the iodine and add one drop of it to each food sample. Do not touch the food with the eye dropper so you don't mix one food with the next.
4. A blue/black or purple color shows that starch is present. The iodine alone is a rusty orange color. Dark-colored foods such as grape jelly or chocolate will be harder to analyze. Figure out a way to make sure you judge these tests correctly.
5. Record your results on the record sheet by writing in a plus (+) (a positive test) for food where starch was present and a minus (–) (a negative test) where starch was not found. If you are not sure how to rate a test, record a question mark.
6. How did your predictions compare with your test results?

WARNING:

Iodine is poisonous. Do not eat any of the tested samples. Wipe up spills and put all samples in the trash when you are finished.

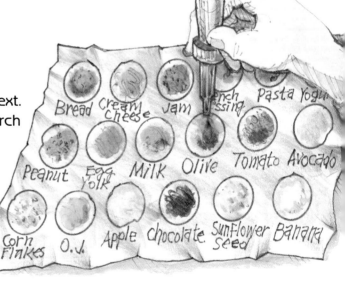

Vocabulary and concepts appear on page 178.

Observations, Ideas, Sketches, Questions

Food Item Tested	Prediction (circle what you think the food contains)	Fat Test Results (+ or −)	Starch Test Results (+ or −)
1.	Fat Starch		
2.	Fat Starch		
3.	Fat Starch		
4.	Fat Starch		
5.	Fat Starch		
6.	Fat Starch		
7.	Fat Starch		
8.	Fat Starch		
9.	Fat Starch		
10.	Fat Starch		
11.	Fat Starch		
12.	Fat Starch		
13.	Fat Starch		
14.	Fat Starch		
15.	Fat Starch		
16.	Fat Starch		
17.	Fat Starch		
18.	Fat Starch		
19.	Fat Starch		
20.	Fat Starch		

Notes:

"What you hear you forget, what you see you remember, what you do you know."
—Chinese Proverb

10. Snap! Crackle! Popcorn!

Food Science / Snap! Crackle! Popcorn!

The food inside corn kernels is there to feed baby corn plants, not people. But we find it so tasty that Americans eat an average of 16 gallons of popped corn per year*. Popcorn is fun to make as well as to eat. In this activity, you will make several batches as you investigate how the water content of corn kernels affects kernel popping. What, there's water in popcorn? Read on!

Materials

- Popcorn (loose kernels, not the kind for microwaving)
- 2 cups or small bowls
- An oven
- 1 popcorn popper OR a stove burner and a small pot with properly fitting lid (A glass pot and/or glass lid is nice for watching the kernels pop.)
- Paper towels
- 1 cookie sheet with sides OR cake pan
- 1 measuring cup
- Cooking oil, unless using an air popper
  (Do *not* substitute butter; it burns.)
- 1 stopwatch
- Salt
- Foil, paper, glue, optional
- Balance from "Balance for Small Weights" (see page 109), optional

Background

Popcorn kernels may not look like they contain any water, but they do. It's hiding throughout the starchy insides. When heated, the pressure inside the kernel increases to about 325°F (163°C), the water turns to steam and the force of this internal pressure explodes the seed coat. When the coat breaks, the steam expands further, creating zillions of tiny gas bubbles that puff up the starch. You are going to find the effects of increasing and decreasing the amount of water in corn kernels. This will be done by drying one batch, soaking another, popping them, and comparing each to kernels popped straight from the bag.

Warning:

Parent supervision of this activity is necessary for safety. Popping corn requires the use of an electric popper or the stove. If a pot is used, the lid must fit properly so that hot oil does not splatter out. The steam produced during popping is also dangerous.

"The man who has no imagination has no wings."
—Muhammad Ali

*From The Popcorn Board, Chicago, Illinois.

I. Popcorn Experiment

Total activity time for parts A, B, and C: 50–100 minutes (plus 2 1/4 hours for drying and soaking)

A. Prepare Three Sets of Popcorn Kernels

Activity time: about 10 minutes (plus 2 1/4 hours for drying and soaking)

Set #1: Dried

1. Preheat your oven to 200°F (92°C)
2. Count out 100 kernels of popping corn onto the cookie sheet.
3. Place the cookie sheet in the oven and dry the kernels for 2 hours.
4. Remove the cookie sheet from the oven and let it cool.

Set #2: Soaked

1. Count out 100 kernels of popping corn and place them in a cup.
2. Add water so that it reaches above the level of the kernels. Set the cup aside to soak the corn for 2 hours.
3. Pour off excess water and dry the kernels with paper towels.
4. Let the kernels sit out on dry toweling about 15 minutes to dry completely on the *outside*.

Set #3: No Pretreatment

1. When you are ready to pop them, count out 100 kernels of popping corn. This is your control group.

WARNING: It is dangerous to add water or anything wet to hot oil.

First pour off water

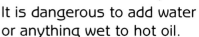
Then dry, and dry some more

"Wonder is the beginning of wisdom." —Greek proverb

Food Science / Snap! Crackle! Popcorn!

B. Pop the Corn

Activity time: 20–30 minutes

1. Read ahead and look over the record sheet. If you want to time how long it takes for the kernels to begin popping and/or how long they take to pop, have your stopwatch ready.
2. Pop the untreated batch of kernels first.
3. If you have a popcorn popper, follow the directions for its use.
4. If you are using a pot, put in 1 tablespoon oil or just enough to cover the bottom.
5. Add the corn (100 kernels) and swirl the oil over and around them.
6. Cover the pot and put it on the stove over *low* heat. Wait until you hear the popping. Don't touch the pot until step 8.
7. Keep listening. The popping will speed up, then slow down. When the popping slows so that 2 seconds go by without a pop, remove the pot from the heat. Don't wait longer or the popcorn will burn. Set the pot on a cool burner or hot pad, not directly on the countertop.
8. Wait at least 30 seconds before opening the pot. Sometimes a few more kernels will pop.
9. Open the pot cautiously, away from your face.
10. Repeat the popping process for each batch of corn you prepared. If any of the corn burns, scrub the pot before doing the next batch. Just be sure to dry it completely before you use it again.

Teacher Tip

Dedicate a set of new 400 mL beakers to "edible use only." One tablespoon (15 mL), or 100 kernels, of corn and two teaspoons (10 mL) of oil are perfect amounts to pop a beakerful of regular popping corn on a hotplate. 400°F to 450°F (204°C to 232°C) is the temperature range needed. Cover beaker with foil before heating. Poke a few holes in the foil. Kids enjoy watching the corn pop through the glass. Small lab groups can make their own. Goggles and beaker tongs are needed for safety. Salt and napkins recommended! Warning: Do not use this method for soaked corn as it pops too violently and will push off the foil. Prepare soaked corn in a pot with a lid.

WARNING:

Do not lift the lid! When kernels pop uncovered, they shoot out along with burning hot oil.

STEAM WARNING:

Steam will escape and it can burn you!

WARNING:

The soaked corn will pop violently, so the lid must fit the pot properly.

Tara: Do you know the name of today's greatest labor-saving device?
Sara: No, what is it?.
Tara: Tomorrow!

C. Collect Data and Compare the Results

Activity time: 20–60 minutes

1. How will you compare the results of the three sets of kernels? There are many ways to measure them, and you can choose all or some of them. You can compare the number of kernels that popped and how long it took them to begin and/or finish popping. You can describe the shapes of the popped kernels and use sketches to illustrate your words. You can compare the total volumes of each popped sample. You can randomly (with eyes closed) pick out ten popped kernels from each group and line them up, measuring the length. If you glue these to paper in three columns, you can create a nice three-dimensional bar graph showing sizes after popping. Oil from the popcorn will soak into paper, so glue strips of foil to your paper and then glue the popcorn to those.

2. Record your data and observations on the record sheet.

3. Two other bar graphs could be done, one to show the comparison of popping times and another to show how the final volumes of popped corn compared. The Graphing Appendix on page 204 can help you set up bar graphs.

4. Think of any sources of error you might have had and list them on the record sheet. Any condition that differed from one popcorn group to the next, other than water content, should be listed.

5. Write a conclusion about the effects of the different treatments on the popping of popcorn.

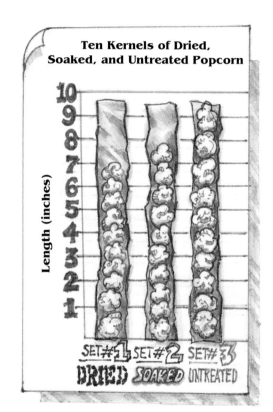

II. Other Investigation Ideas

Activity time: will vary

- Do a taste test of popcorn that was salted before cooking compared to popcorn salted after cooking. Is there any difference in the flavor or tenderness of the puffs?
- Find the percent of water in popping corn fresh from the package. Make a prediction before you begin. (See "Water in Ketchup?" on page 114.) The idea is that you will find and compare the weight of water lost by drying to the weight of the original corn.

    weight of water lost ÷ weight of original corn × 100 = percent water

Use your balance for small weights (page 109) or other balance. With the balance for small weights, collect corn kernels on one side that balance with your test sample of

100 kernels. Put these comparison kernels in a plastic bag to use again later. Dry the test sample of unpopped corn at 200°F (93°C) overnight. Weigh again. You can calculate, or at least approximate, the percent of the corn that is water by seeing how many kernels you need to remove from the comparison sample to balance with the dried sample. If you have to remove one kernel from your equivalent sample to balance, then the weight went down by one out of 100 or 1%. If it takes between five and six kernels to balance, the water content was between 5% and 6%.

- Find the effect of different lengths of drying time on the number of kernels popped or the popped volume. For example, count out five or six batches and dry each for a different length of time, such as 15 minutes, 30 minutes, 45 minutes, and so on. Create a line graph of results, using volume or number popped vs. time dried. See the Graphing Appendix on page 204 for help with setting up the line graph.
- Try sprouting popcorn. Soak some kernels in a cup overnight, pour out the extra water, and cover the cup with a wet cloth or paper towel held in place with a rubber band. Keep the kernels moist by rinsing them twice a day. Examine the roots and leaves that grow. Or, after soaking overnight in water, plant them in moist soil 1/2 inch (1 cm) below the surface. Place in a sunny spot. They will grow!

Vocabulary and concepts appear on page 179.

Snap! Crackle! Popcorn! Experiment

How Does Changing Kernel Water Content Affect Popping?

Purpose:
The purpose of this investigation is to find out

Prediction:

Data Table:
Choose the types of data you want to collect

	Untreated Kernels	**Dried Kernels**	**Soaked Kernels**
Time to Begin Popping (sec)			
How Long Popping Lasts (sec)			
Number Popped			
Volume of Total Sample After Popping (cups or milliliters)			
Length of 10 Popped Kernels (inches or centimeters)			

Snap! Crackle! Popcorn! / Record Sheet

	Untreated Kernels	**Dried Kernels**	**Soaked Kernels**
Sketches of Sample Kernels			
Description of Popped Kernels			

Sources of Error:

_____
_____
_____

Conclusion:

_____
_____
_____
_____
_____

11. Boogie Woogie Butter

It is fascinating to watch cream transform into butter! Just when you think it can't possibly work, solid, light-yellow butter will appear before your eyes. Try it! Have some fresh bread or crackers ready to enjoy with your real homemade butter. To investigate further, you can make whipped cream and design a butter-making experiment.

Did you hear about the man who walked down the street and turned into a bakery?

76

Create Your Own / Boogie Woogie Butter

Materials for Butter

1/2 to 1 cup (120 to 240 mL) heavy cream or whipping cream
1 measuring cup
1 jar with a screw-on lid that is watertight and at least big enough to be only half-full when cream is added (1 to 2 cups [240 to 480 mL] or bigger)
1 cereal-size bowl
Salt, optional
1 spoon or spatula
Music with a zippy beat
Plastic wrap

Materials for Whipped Cream

1/2 pint (1 cup) (240 mL) heavy cream or whipping cream
Electric mixer
Mixing bowl
Sugar

Teacher Tip

To complete butter making, eating, and cleanup in one 40-minute period, discuss the instructions and have the students prepare the jars (steps 2–4) the day before the butter-making day. They can do the rinsing and pressing of their butter (step 10) in their jars. Tongue depressor sticks may be used instead of spoons.

I. Making Butter

Activity time: about 45 minutes

1. Put the bowl into the freezer to chill. You will use it a little later.
2. Test to see that the jar seals well by adding water, tightening the lid, and shaking the jar. If water leaks out, try a different jar. Sketch and describe your chosen jar on the record sheet.
3. Fill your jar to the brim with water to see what volume (space) your jar holds. Measure the water as you pour it out with your measuring cup (or graduated cylinder) and record. You may have to fill the measuring cup several times. Add these volumes together for your total volume.
4. Wash the jar and lid with soap and rinse well. Dry your jar.
5. Measure the amount of cream you have chosen, pour it into the empty jar, and tighten the lid. You don't want to use so much cream that your jar is more than half full. Record the volume of cream you used on the record sheet.

6. Turn on some music with a good beat. Butter is made by shaking the jar, so go to it with energy. Up, down, in, out, back and forth. This will be a 5-minute to 20-minute workout. Boogie down!

7. You should know that the cream will go through three stages on its way to becoming butter. It's OK to look into the jar at any time. In the first stage, you feel the runny cream moving about as you shake the container. A little later, the cream will have thickened so that it feels like it doesn't move at all when you shake it. If you look inside the jar at this point, the cream looks like dense whipped cream and is all one color. Don't quit as you are still making progress!

8. After a while, the contents of the jar will, once again, start to slosh. It will sound like water is inside. This change happens suddenly. Look inside. If you have one blob of what looks like butter and it is surrounded by runny milk, you are done! If the butter is not in one lump, shake another minute as the butter finishes separating from the milk.

9. When the butter is ready, it forms a single, light-yellow blob surrounded by runny white milk.

10. Pour the milk out into a cup for drinking or to use in baking later. Put the butter into your chilled bowl from the freezer and press it with a spatula or spoon to get all the milk out. Rinse the butter well with cold water, pressing it some more. You want to remove all the milk because the milk will go bad in a few days but the butter will last for a few weeks.

Create Your Own / Boogie Woogie Butter

11. Try your fresh butter. It will be soft and spreadable. If you prefer lightly salted butter, sprinkle it with a little salt and stir.
12. Before you refrigerate what's left, you might want to shape it. You can press it into a small dish and use kitchen utensils to draw your initials or make a design on top. When cooled, the butter will harden into this shape.
13. Cover with plastic wrap and refrigerate.

II. Making Whipped Cream

Activity time: 10 minutes, plus 15 minutes in advance to chill bowl and beaters

If you've never made your own whipped cream, be sure to try this. You start with the same cream used to make butter. It tastes *so* much better than store-bought "whipped topping."

1. Put bowl and beaters into the freezer for 15 minutes ahead of time. Be sure the cream is cold.
2. Pour the cream into your chilled mixing bowl.
3. Whip at high speed with the electric mixer. When it's starting to look like whipped cream, whip in sugar until it tastes sweet enough. Start with 1 rounded tablespoon (15 mL) for one cup (240 mL) of cream. Add sugar until it tastes right to you.
4. Whip the cream until peaks hold their shape. To form a peak, turn the mixer off and lift it up. If the cream under the lifted beaters takes the shape of a chocolate kiss and doesn't flatten out, the whipped cream is ready.
5. Try the whipped cream on fruit such as strawberries or blueberries. Delicious!

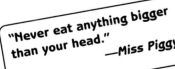

"Never eat anything bigger than your head."
—Miss Piggy

III. Another Way to Make Butter

Activity time: about 20 minutes

If you whip cream too long you end up with butter. You can make butter this way, just don't add sugar. Keep whipping and eventually you will see the cream divide into two parts, solid butter and runny milk. Stop whipping and collect the butter together with a spoon. Continue with steps 10–13 in the directions for making butter.

IV. Butter Experiments

Many factors can affect how long the butter-making process takes. Design an experiment to test the effect of one of these variables on the time it takes to make butter by shaking:

- Size of jar
- Volume of cream
- Type of cream (heavy vs. light)
- Temperature of the cream
- Shaking method (up and down vs. swirling vs. figure 8)
- Presence of a marble to agitate the cream (If you use a marble, be sure to use a plastic jar as a glass one might crack.)

1. Pick the variable you want to test. Pick just one!
2. Within the variable, pick two settings. For example, pick two temperatures, two shaking methods, or a marble vs. no marble.
3. Prepare at least two jars to make butter, at least one of each setting. Unless the volume of the jar is your chosen variable, use the same kind of jar. More trials (more jars) of each setting would be even better. You can then average the results of the trials for a more reliable conclusion. Decide how many you can do by the limits of your time and budget.
4. Every variable *except* the one you are testing should be the same in all jars. So if you are testing the effect of having a marble, half of your jars will have a marble and half will not. All the jars will be the same; the volume, type, and temperature of the cream will be the same, and the shaking method will be the same.
5. Record the time you begin shaking. Record the time each jar finishes. With more than two jars, you will need to find a way to get all the jars shaken the same way. Find partners to mirror your motions or connect your jars together well with duct tape. Can you invent a shaking gizmo that will hold several jars?

"Whatever you can do, or dream you can, begin it. Boldness has genius, power, and magic in it."
—Johann W. Von Goethe

Vocabulary and concepts appear on page 180.

Observations, Ideas, Sketches, Questions

Making Butter:
Sketch of Butter-Making Jar:

Volume of Jar: _____
Volume of cream: _____

Time shaking begins: _____
Time butter forms: _____

How long it took for the butter to form: _____ (Subtract the time shaking begins from the time butter forms to tell how long it took in minutes for the butter to form)

Observations:

Whipping Cream Observations:

Butter-Making Experiment

Purpose:
The purpose of the butter-making experiment is to find out the effect of _____ on the time it takes to make butter in a jar.
(your test variable)

Prediction:
_____
_____

Procedure:
List your steps with enough detail that someone else could repeat exactly what you did.
1.
2.
3.
4.
5.

Data:
This section tells your results in a table. A quick glance at the table should tell you exactly what happened. In this case, you will give the number of minutes of shaking that each jar needed. For example,

Treatment	Starting Time	Ending Time	Butter-Making Time (minutes)
No marble	3:25 P.M.	3:43 P.M.	18
With marble	3:25 P.M.	3:34 P.M.	9

Treatment	Starting Time	Ending Time	Butter-Making Time (minutes)

Conclusion:
Write a statement that sums up your results in sentence form. It directly fulfills your purpose. For example, for the invented data here, you would write "Adding a marble to the jar cuts the butter-making time in half."
_____
_____
_____

12. Yogurt, Please

Cows make milk, but cows do not turn milk into yogurt. The creatures that do that are not mammals, they don't have legs, and they don't even have eyes or a brain. They are bacteria, tiny living chemical factories that change sweet, liquid milk into thick, tangy yogurt. The bacteria need a work shift of 6 hours or more to finish the job, so you might want to make your yogurt one day and eat it the next.

Materials

- 2 cups (480 mL) milk (of whatever fat content you prefer: whole, 2%, skim)
- 1/2 cup (120 mL) instant nonfat dry milk, optional (increases firmness)
- 1/2 cup (120 mL) plain yogurt (completely plain with *living* cultures), at room temperature to use as "starter"
- Cooking pan
- Stove
- Containers for the yogurt such as foil-covered cups or plastic containers with lids
- Candy thermometer
- Spoon
- Medium-size cooler (a double cardboard box will also work)
- 3 or 4 quarts (or liters) of warm water in containers such as soft drink bottles
- Sugar, honey, or fruit preserves
- Fresh fruit, optional
- Vanilla, optional

83

I. Make Yogurt

Activity time: about 1 hour

The big picture is that you will heat milk to 160°F (71°C) to kill unwanted bacteria, cool it to 110°F (43°C), which is the perfect temperature for the good bacteria, and then keep it cozy at that temperature while the live cultures multiply and turn the prepared milk into yogurt.

1. Decide what container to make the yogurt in. You can use cups for single serving portions, or you can use a larger container to hold it all. Have the container(s) ready.
2. Prepare a warm nest for the container(s). Fill plastic soft drink bottles or any other containers with hot tap water at about 120°F (49°C). Put them into the cooler (it's a "warmer" now!) leaving room for the container(s) that will hold the prepared milk.
3. Measure the milk into the pan. If you are adding dry milk, do so now and stir it in. Heat the milk on the stove over medium heat. **Warning:** Hot stove! Careful! Use the candy thermometer to keep track of the temperature. When it reaches 160°F (71°C), turn off the heat and move the pan to a cooler spot.
4. Keep watch on the thermometer as the milk cools. This takes a while, so find something to read to pass the time. About every minute, stir the milk and check its temperature. When the milk cools to 110°F (43°C) add a 1/2 cup (120 mL) of yogurt and stir it in gently.
5. Pour the 110°F (43°C) milk into the container(s) you have chosen and cover with foil or lid(s).
6. Place container(s) into the cooler. The more warm stuff you have in the cooler, the longer the milk will stay at the right temperature.
7. Record your observations so far under "Day 1" on the record sheet.
8. Check in 6 to 18 hours. The yogurt is ready if it stays in one custard-like form when you tip the container to the side. It's OK if there's some clear liquid at the top. It may not be quite as thick as what you get in the store.

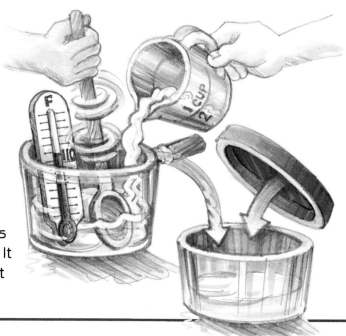

Create Your Own / Yogurt, Please

9. Record your observations on the record sheet under "Day 2."
10. While it's fine to eat it now, you might prefer it cold. If so, refrigerate it for a few hours before eating.
11. Most of the yogurt you've had before was probably sweetened and flavored. Add sugar or honey to make it sweet. A touch of vanilla is tasty. Fruit preserves add both fruit and sweetness, so skip the sugar if you use them. You may or may not want sugar along with fresh fruit. Refrigerate uneaten yogurt.

Teacher Tip

Keep a case of new 400-mL beakers set aside just for edible labs. Using hotplates, students can use the beakers to make their own yogurt serving using 1/2 cup milk and a tablespoon of yogurt instead of the amounts shown here. Stir with craft sticks. The labeled, foil-covered beakers can be arranged in the cooler in two layers with a sheet of cardboard between layers. If it is not convenient to have students do the heating and cooling steps, you can do that part for all; then each student can stir the yogurt into his or her milk portion in a paper cup, cover it, and during class the next day, check it, and eat. If possible, refrigerate the yogurt a few hours before eating.

II. Yogurt Investigations

Q: Why did the bacterium cross the microscope?
A: To get to the other slide.

Activity time: about 1 hour

Think about all the conditions (variables) of the yogurt-making process. To design your own experiment, pick one condition to change. Keep all other conditions the same. Compare the results with and without the change. For example,

- Find the effect of the powdered milk by making two yogurt batches, one with and one without powdered milk.
- Find the effect of the type of milk used. Compare yogurt made from skim milk to yogurt made from whole milk.

- Try to discover the temperature at which the good bacteria are killed. It is somewhere above 110°F (43°C), but where? Start the yogurt according to the recipe, but don't cool the milk as much. Put 2 tablespoons (30 mL) of room temperature yogurt starter into each of three containers labeled "122," "118," and "114." When the milk cools to 122°F (50°C), pour a third of the milk into that container and stir. (Got an assistant to help you?) Pour half of what's left into the second container when the milk cools to 118°F (48°C), and pour off the last portion at 114°F (46°C). After stirring, cover and put all into the prepared cooler. Once you find out the results of these tests, you can narrow your answer down further by . . . hey, you figure it out!

III. Insulation Investigations

Design an experiment to find out how different materials compare as insulators. Try different containers and/or filler materials around identical bottles of hot water. Record the starting water temperature. After a set amount of time, remove the bottles and compare the temperatures. This would be fun to do as a competition with a few friends. Each person starts with a cardboard box and identical bottles of hot water but chooses his or her own materials for insulation. Whose water is the hottest after 30 or 40 minutes?

"I base my fashion taste on what doesn't itch."
—Gilda Radner

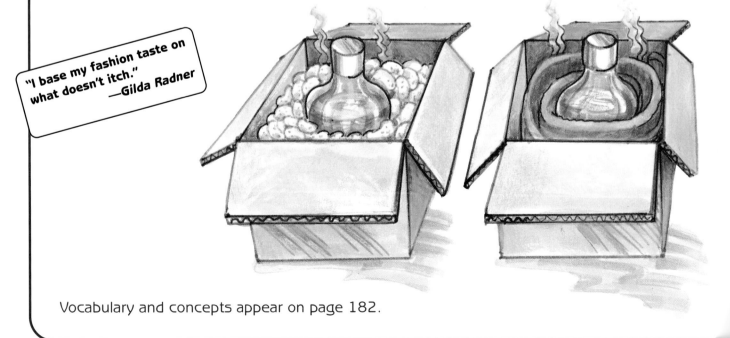

Vocabulary and concepts appear on page 182.

Yogurt, Please / Record Sheet

Observations, Ideas, Sketches, Questions

Day 1 Observations While Preparing Milk:

Day 2 Observations of Yogurt:

Investigation Records:
(See page 88 for full "Yogurt, Please Experiment" form)

Father: I hear you skipped school to play football.
Son: No, I didn't, and I have the fish to prove it!

Yogurt, Please Experiment

Purpose:
The purpose of this investigation is to find out _____
_____
_____

Prediction:
_____
_____

Data and Observations:

Sources of Error:
_____
_____

Conclusion:
_____
_____
_____

13. Sprout Jungle

What vegetable can be grown any time of the year, is ready for harvest in less than a week, needs no soil or sun, is high in protein and vitamin C, and needs no chopping or other preparation?

You've got it . . . sprouts!

A sprout is a baby plant—roots, stem, leaves and all—just a few days out of its seed coat. You have seen bean sprouts on salad bars and in stir-fry dishes. If you don't like eating them, watch them grow, and then feed them to your health-conscious friends.

Judy: Why did the tap dancer retire?
Rudy: He kept falling into the sink.

Materials for Making Sprouts

1 quart-size jar
1 tablespoon (15 mL) mung bean seeds (health food stores and some grocery stores carry them)
1 tablespoon (15 mL) household chlorine bleach
Cheesecloth, a handkerchief, or other piece of thin, clean cloth big enough to cover the mouth of the jar
1 rubber band
Magnifier, optional
1 strainer

Warning:
Bleach will sting eyes and skin and discolor clothes. Pour it carefully over the sink.

Warning:
Don't use seeds packaged for planting as they may be treated with chemicals that should not be eaten.

> "If stupidity got us into this mess, then why can't it get us out?" —Will Rogers

Materials for Charting Progress

1 ruler
Scale or balance sensitive enough for small weights like seeds (See "Balance for Small Weights," page 109 for a fine balance you make yourself.)

I. Making Sprouts

Activity time: 10 minutes the first day plus 15 minutes wait time; 5–20 minutes a day for the next four days, depending on your level of record keeping

The big picture is that you will clean and soak the seeds, and then put them in a dark place to sprout, rinsing daily to keep them moist. If you plan to chart the progress of your sprouts as they grow, read "Charting Progress" on page 91 now.

1. Put 1 tablespoon (15 mL) of seeds into the jar and swirl with warm water. Pour the water off without losing the seeds. Repeat if the rinse water is not clear.
2. Fill the jar one-quarter full with warm water and add the chlorine bleach. Let sit about 15 minutes. The bleach will kill any fungi, yeast, or bacteria that may be clinging to the seeds.
3. Pour off the bleach solution and rinse the seeds again.
4. Add clean water to fill the jar half way. Cover your jar with the cloth and use the rubber band to hold it on.

5. Now let the seeds soak overnight or for about 8 hours. During this time the seeds will absorb water and the sprouting process will be triggered.
6. After the soaking, pour off the water. You can pour the water right through the cloth, letting the cloth hold back the seeds.
7. The seeds and the inside of the jar are now wet, but there is no standing water. Set your jar in a cabinet or drawer that is not often used so that the sprouting occurs in darkness.
8. Rinse the seeds once (in a humid climate) or twice (in a dry climate) every day, pouring off the extra water each time.
9. About the fourth day, your sprouts are ready. Take off the cloth, put them in a strainer and wash them again to get rid of some of the dark green seed coats that were shed when the seeds sprouted. Time to eat! Mung bean sprouts are good in salads, on sandwiches, and in stir-fry dishes.
10. Keep leftover sprouts loosely covered in the refrigerator, and eat them within a week.

II. Charting Progress

You might want to keep track of the changes your seeds make each day. You can observe carefully with your senses, make sketches, measure their weight and length, and write down all the information you collect. If this is a science fair project, you could then create one-page displays of each day's changes.

You will notice that the seeds don't all look alike and neither do the sprouts. So when you make your observations and measurements be sure to look at several. That way you learn something about the trend of what's going on, that is, what's typical, what's middle-of-the-road. Where possible, you should find averages such as average length or average weight. On the record sheet, write down *everything you measure* as well as the averages you calculate. If you don't, errors can't be found or fixed. (And even *you* make mistakes now and then!)

Sketch On the record sheet, sketch three to five seeds or sprouts each day.

Measure Length To measure seeds, line up five seeds end-to-end, record the total length, then divide that length by 5 and record. For sprouts, gently stretch without breaking them one at a time along a ruler to measure length. Add the five lengths and divide by 5 to get the average or mean length.

Measure Weight *If you have a manufactured school or kitchen scale*, begin this step on Day Zero. Count out ten seeds. Dry them gently with paper toweling and weigh them together. Record. Divide that weight by 10 to find the average or mean weight of one, and record.

If you are using your homemade balance, begin the weight tracking on Day One. Count out ten growing seeds/sprouts. Dry them gently but well with paper toweling and place them together in one of the baskets. Add original dry seeds to the other side until they balance. Count the dry seeds you needed and record. Divide that weight by 10 to get the average weight of one growing seed. Your answer will tell you how many times heftier a soaked seed or sprout was than the original seed. Let's say your answer is 2.4. That means a growing seed is almost two and a half times the weight of a dry seed. Where did it get this extra weight?

To convert a weight in ounces to a mass in grams, use the appendix on page 210.

III. Sprout Investigations

Activity time: will vary

Design your own experiments by changing one condition (variable) between two groups of sprouts and then making comparisons (as in "Charting Progress" on page 91) between the two groups. For example,

- Does it seem strange that sprouts grow in the dark? Start two jars of sprouts, one kept in a dark cabinet and one kept on the kitchen counter in the light. How do they compare in growth and taste? How do they compare when you extend the growing time to 10 or 15 days?
- Alfalfa, radish, onion, wheat, and soybean seeds are also used to make sprouts. They are available in health food stores. Try sprouting some of these and see how they compare. The alfalfa sprouts are popular and the seeds are easy to find. The main difference in working with them is that they are tiny.
- What about the effect of temperature? Grow two, three, or four jars of sprouts at temperatures that differ by at least a few degrees. Use a thermometer to measure the temperature in each location. One location could be the refrigerator.
- Other possible variables include the color of light (wrap jars with colored cellophane) and the presence of other substances such as salt or baking soda in the rinse water.

Start planning!

Vocabulary and concepts appear on page 183.

Observations, Ideas, Sketches, Questions

Day Zero (before soaking has begun)		
Sketches and Observations:	**Length Data:**	**Weight Data:**
		(On the homemade balance, weight data won't mean much until Day One. It will be 1 times weight of dry seed.)

Day One Date:		
Sketches and Observations:	**Length Data:**	**Weight Data:**

"In this life, we cannot do great things. We can only do small things with great love."
—Mother Theresa

Sprout Jungle / Record Sheet

Day Two **Date:**		
Sketches and Observations:	**Length Data:**	**Weight Data:**

Day Three **Date:**		
Sketches and Observations:	**Length Data:**	**Weight Data:**

Day Four **Date:**		
Sketches and Observations:	**Length Data:**	**Weight Data:**

14. Herb Garden

It's fun to watch seeds turn into delicate plants that poke their way through the soil and reach for the sun. It's even more fun if you get to eat them when they get bigger. Herbs are plants that have uses as medicines or food flavorings. Try out your green thumb with a few tasty herbs and watch the transformations. Water is the key!

Materials

Available at a garden store:
  Choose a few of the following herb seeds: arugula, basil, cilantro (also called coriander), onion, parsley, peppermint
  Potting soil
Containers such as plastic "flats," flower pots, coffee cans, or plastic containers (poke a drainage hole in the bottom of them with a nail, if possible)

Small rocks
Plates or trays for under the containers
1 light-colored plate
1 clean, water-filled spray bottle
1 ruler
1 marker
Plastic wrap
Liquid fertilizer
Space in front of sunny windows

I. Getting Started

Activity time: about 30 minutes

1. Warning! You are bringing life into the world. Once you begin, you become the parent of living things that require care for weeks and months. No diapers to change, fortunately, but you can't forget about them for a few weeks and expect them to survive. If you think you've got what it takes to be a plant parent, carry on!
2. See how much space is available for you to use in front of sunny windows in your home. You will start your plants there. If possible, you will later move them to a deck, patio, or outdoor garden.
3. Collect containers that will fit in your space. You may need a shelf or stand to hold them.
4. See what herbs are available at the garden store and choose a few that you and your family like. Three or four types are plenty. Select short or dwarf varieties, where possible. (After herbs get a start, they grow best outdoors, but all the ones listed here can continue indoors if it's not the right season or you don't have an outdoor space you can use.)

Teacher: You aren't paying attention, Bob. Are you having trouble hearing?
Bob: No, I'm having trouble listening!

Create Your Own / Herb Garden

5. Put a rock over the drainage holes in your containers before adding soil. Don't use soil directly from the outdoors because you might bring unwanted little critters into the house. Potting soil is sterile; it has no weed seeds and no critters. Add the potting soil outdoors or over a sink, since it can be messy. Fill your containers almost to the top with soil, pat it down, and add water until all the soil is moist but not soggy.

6. Seeds that are slow to sprout (check the package) can be sped up by soaking overnight in water. Just put water in a cup and add the number of seeds you think you will use. Use a different cup for each kind of seed and label them. One day later, drain the water off the seeds, put the seeds on paper towels, and then plant as described in step 7. This step (6) is optional and should not be done with basil or arugula because they get gooey if soaked.

7. Put seeds onto a light-colored plate. Peppermint, parsley, and basil seeds are very tiny and so are tricky to handle. For those types, pick up seeds by pressing a finger on them, then, with a rolling pinch, sprinkle them onto the soil. Repeat as needed to cover the surface, spacing the seeds about $1\frac{1}{2}$ to 2 inches (4 to 5 cm) apart. The bigger seeds can be picked up and placed with the same $1\frac{1}{2}$ to 2 inch (4 to 5 cm) spacing. You will have seeds of each kind left over.

8. Find the planting depth recommended on each seed package. Use a ruler and a marker to mark on the pot how much soil is needed to cover the seeds. Add soil and pat it down so that it reaches the line. Use the spray bottle to moisten the soil.

9. Label each pot with the name of the herb. Be sure to place your pots on plates or trays so that you don't cause water damage to your windowsill or furniture.

10. To keep in warmth and moisture during germination, cover your pots with plastic wrap. Take off the wrap after your plants appear.

11. Peppermint, onion, and arugula will germinate in 1 to 2 weeks. Parsley, basil, and cilantro will germinate in 2 to 3 weeks. Read the back of the seed pack for more detail. Gardening takes patience, doesn't it?

"A good laugh is sunshine in the house."
—William M. Thackeray

II. Raising Happy Plants

Activity time: 5 minutes every other day

Plants are pretty easy to please. They don't need food because they make their own (that's what photosynthesis is all about). They don't need to go to the bathroom because their only wastes go into the air (such as oxygen). But they are fussy about getting plenty of sun and the right amount of water. And they don't grow well when they are crowded.

Sun In the northern hemisphere, choose windows on the east, south, or west sides of your home. North-facing windows don't provide enough sun. In the southern hemisphere, it's the south side that isn't sunny enough, so choose east, north, or west. Plant lights can help your indoor plants get enough light.

Water Check the soil daily. If it feels moist, don't add water. If it feels dry, spray it well. If you see signs of wilting, you waited too long to water, but most plants will revive from an occasional wilt.

Thinning Once your plants are a few inches tall, you need to thin them out, that is, select which ones will continue to grow. Snip off any weak-looking seedlings and others until the remaining plants are evenly spread out and about 3 to 4 inches (7 to 10 cm) apart. You can eat the snippings and get your first taste of the young herbs.

Keep the strongest plants.

Fertilizing While plants make their own food using sunlight, they must get their minerals from the soil. Fertilizer replaces minerals that get used up. Use a liquid fertilizer according to directions for indoor plants.

III. Moving Plants Outside

Activity time: will vary

Once your plants have a good start, you might want to move them to a porch, patio, or outdoor garden where their growth can speed up. If so, make the move gradually so that they aren't shocked by the change in their conditions. This process is called hardening off. Take them outdoors a few hours one day, then a little longer each day for several days. Depending on humidity and rainfall, you may need to water more or less when they are outdoors. To transplant an herb into the ground from a pot, first dig your holes in the soil. Run a knife along the wall of the pot around the soil and try to remove the soil and plants as one unit. Then gently separate the individual plants with as little tearing of the delicate roots as possible, and place each one in its new hole. Tamp the soil down around the plants for support.

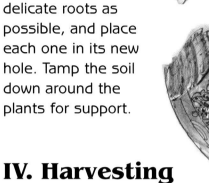

IV. Harvesting

Activity time: a few minutes

Once **arugula**, **cilantro**, and **parsley** plants have about six leaves, pick the outer leaves to eat and the plant will keep producing from the center.

Basil and **peppermint** can be encouraged to grow bushier if you occasionally snip off the growing tip. Individual leaves can be pinched off for use.

Onions are pulled out whole as green onions.

V. Eating

Arugula This herb gives punch to salads. The taste of arugula alone is too strong for most people, but a little mixed in with lettuce is delicious. Tear the leaves into small pieces and mix with lettuce.

Basil Cook whole basil leaves on pizza. Or chop it to season spaghetti sauce and other Italian dishes.

Cilantro Chop the leaves and sprinkle them on salsa, salads, soups, sandwiches, tacos, and Thai and Chinese dishes. People tend to love it or hate it.

Green Onion Trim off the roots and eat as is or slice into small discs and add to salads, soups, potatoes, and meat dishes.

Parsley Add to salads and sandwiches. Use it to decorate platters of food.

Peppermint Add whole leaves to flavor iced or hot tea or experiment with it in other drinks. Put a little on sandwiches of sliced cucumber and cream cheese. Chew leaves as a breath freshener.

VI. Investigation Ideas

"Patience! The windmill never strays in search of the wind."
—Andy J. Sklivis

Activity time: will vary

- Find out how much difference soaking the seeds makes in germination time for different types of herbs.
- Compare the size and color of herbs germinated in the dark with those germinated in the light.
- Find the water percentage of different types of herbs by adjusting the method described in "Water in Ketchup?" on page 114.
- Compare the taste of an herb before and after the plant has flowered. You need to have plants at different stages of growth so you can taste them at the same time.

Vocabulary and concepts appear on page 184.

Observations, Ideas, Sketches, Questions

Herb Garden Observations

Name of Herb			
Seeds Soaked?			
Date Planted			
Date Seedlings First Seen:			
Notes Week One			
Notes Week Two			

Herb Garden / Record Sheet

Notes Week Three			
Notes Week Four			
Notes Week Five			
Notes Week Six			

15. Kitchen Compost

Q: What do you get when you pour steaming water down a rabbit hole?
A: Hot, cross bunnies!

The natural world recycles everything. When living things die, their bodies rot back into the soil. If they did not, there would be heaps of dead stuff all around. Imagine! Active decay isn't pretty, but it takes care of business. When decay is complete, the result is black, clean-smelling stuff called compost or humus, and it is good for plants. Watch the natural recycling process in this activity and then treat your outdoor plants to the rich, finished compost you make.

103

Materials

- 2 clean 5-gallon (18-L) plastic buckets with lids (Restaurants and cafeterias often have these to give away. Hardware stores sell them.)
- 1 Drill and adult to help
- 3/8- to 1/2-inch (1- to 1.3-cm) drill bit
- 1 medium to heavy-duty sheet of plastic about 5 feet by 5 feet (1.5 m by 1.5 m)
- Natural soil from outdoors (Potting soil from the store doesn't have all the good microbes you need.)
- 1 shovel
- Fruit and vegetable scraps such as carrot peelings, apple cores, and lettuce scraps. Avoid lemon, orange, banana, and grapefruit peels because they are slow to decompose in most soils.)
- 1 trowel for stirring
- Red worms, optional

I. Making and Using the Compost Bin

Activity time: about an hour to prepare the two bins.

You will prepare two bins but keep only one in active use. That way you can pour the compost from one bin to the other to turn it, observe progress, and mix in new fruit and vegetable scraps. Do this over the plastic sheet to make clean up easy.

1. Collect fruit and vegetable scraps for a day or two in a plastic container kept in the refrigerator.
2. Have an adult train you to use the drill safely. Use the drill to put air and drainage holes around each bucket, in two rings of about 12 holes each. One ring should be about 2 inches (5 cm) up from the bottom and the other about 2 inches (5 cm) down from the top.
3. Use the shovel to dig up some soil from outdoors. Add it to one bucket until it is about one-third full, loosening clumps with the trowel. Describe the soil on the record sheet.
4. Now add up to 1 quart (1 liter) of the scraps you have been collecting. Thin scraps such as lettuce and cucumber peelings are excellent.

Note: DON'T USE meats, bones, fats, or oils.

Stir with the trowel to mix the scraps with the soil. Record on the record sheet the materials that you added.

5. Add more soil until the bucket is about half full.
6. The soil should be moist. If it feels dry, add some water and stir. You don't want it too wet, though. It should be as damp as a sponge after it's been squeezed.
7. Set the cover lightly on top of the bucket without sealing and put the bucket away. The temperature should not go below 50°F (10°C). If it stays above this temperature where you live, you can keep the bin outdoors. Otherwise, a basement or garage is a good spot. Talk to your parents about where to keep it.
8. Check your compost bin every day or two and record your observations on the record sheet. "Pour" the compost into the second bucket to see what is there. (If you would rather NOT observe active rotting, wait a week before you check it. If the bin smells bad, stir, cover, and let it sit for a few more days.) Add water, as needed, to keep the soil properly moist. Your compost is ready when no food is visible and it just smells like soil.
9. Add new scraps when the old ones disappear. Turn out part of the compost into the second bucket, breaking up chunks as you go. Mix in the new scraps as you add the rest of the compost.
10. Get your family into the habit of saving fruit and vegetable scraps and recycling them this way. Every month or so you can give some of this rich compost to your outdoor plants. Add more soil from outdoors to your bin to keep it about half full.
11. If you like, turn your compost bin into a worm vacation spot by getting a box of red worms (better than earthworms for this), often sold for fishing bait. Add one or two dozen worms to the bin and make daily observations. See how the worms affect the decomposition process. If you tend to find the worms on top, the soil is too wet. If they are mostly at the bottom, it is too dry.

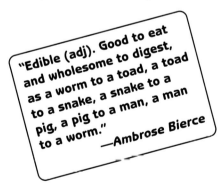

"Edible (adj). Good to eat and wholesome to digest, as a worm to a toad, a toad to a snake, a snake to a pig, a pig to a man, a man to a worm."
—Ambrose Bierce

II. Composting Experiments

There are many variables (factors) in composting that could affect its progress. Design an experiment to test the effect of one of these variables on the time it takes for decay to be complete. Here are some possible variables to test:

- The temperature
- The amount of moisture
- The type of plant scraps
- How finely the plant material is cut
- Turning the soil from bin to bin vs. not turning
- The presence or absence of worms
- The pH (acid/base level) (Soil pH test kits are available at garden stores.)
- What else? What are you curious about?

> " 'Tis better to remain silent and be thought a fool, than to open one's mouth and remove all doubt."
> —Samuel Johnson

1. Pick the variable you want to test. Just one!
2. Within the variable, pick two settings. For example, pick two temperatures at least 5 degrees apart, two different vegetables, or two levels of cutting such as whole carrots vs. carrots grated in a food processor. Label the bins according to your plan.
3. Every variable *except* for the one you are testing should be the same in both bins. So if you are testing, for example, the effect of using whole vs. grated carrots, one bin gets whole carrots and the other gets grated carrots but the soil, moisture, temperature, amount of food, and stirring amount is the same in both.
4. Write a clear statement of your purpose.
5. Write your prediction of what will happen.
6. Write a step-by-step procedure for your experiment with enough detail that someone else could repeat exactly what you did. How will you measure progress? How will you know when "decay is complete"?
7. Record the date you begin the experiment. Decide how frequently you will check on progress.
8. To turn the soil with both bins in use, you will need to empty one onto the plastic sheet, pour the other into the empty one, and then return the soil on the plastic to the second bin. Keep track of which is which by moving labels as needed.
9. Use the data table given or create your own data table with room for dates and regular observations for each bin.
10. Complete the experiment.
11. Create a bar graph to show off your results. The vertical axis will be the number of days to decompose. Along the horizontal axis will be the labels for your two sets of conditions. See the Graphing Appendix on page 204 for more information on making bar graphs.
12. Consider your sources of error. Were there any variables such as temperature or moisture that should have been the same but actually differed from one bin to the other? This is where you honestly review the design of your experiment and note any flaws that could have affected your results.
13. Keeping sources of error in mind, write a statement of conclusion that follows logically from your purpose.

Vocabulary and concepts appear on page 185.

Kitchen Compost / Record Sheet

Observations, Ideas, Sketches, Questions

Description of original soil:
_____
_____

Scrap	Day 1	Day 2	Day 3	Day 4	Day 5	Day 6	Day 7

Scrap	Day ___	Day ___	Day ___	Day ___	Day ___	Day ___	Day ___

Notes:

Description of soil after ___ weeks:
_____
_____

Kitchen Compost Experiment

Purpose:
The purpose of this experiment is to find out the effect of _____ on the time it
(your test variable)
takes for the decay of _____ to be complete in the compost bin.

Prediction:
_____
_____

Procedure:
1.
2.
3.
4.
5.

Data:

Date:	Bin #1 Observations: (Treatment:)	Bin #2 Observations: (Treatment:)

Sources of Error:
_____
_____

Conclusion:
_____
_____

16. Balance for Small Weights

In 15 minutes you can make a balance for weighing small objects that is cheap, easy to use, and incredibly effective. Use it for tracking the development of sprouts ("Sprout Jungle," page 89) and for accurately finding the water content of ketchup ("Water in Ketchup?" page 114). What other uses can you invent?

Create Your Own / Balance for Small Weights

Materials

1 plastic drinking straw
1 ruler
1 straight pin
1 needle, a little fatter than the pin
Clear tape
Thread, heavy-duty, if possible

Two 2 1/2 inch (6.3 cm) paper or foil baking cups, for muffins and cupcakes
Mung bean seeds, lentils, or beads (something small, uniform, and countable)

> "The beginning is the most important part of the work."
> —Plato

Note: For the "Water in Ketchup?" activity you will need *foil* baking cups whether the balance is made with them or not.

Note: Good fine-motor coordination is needed for working with the thread.

I. Making the Balance

Activity time if materials are ready: about 15 minutes

1. Use the ruler to find the center of the straw.
2. Poke the needle at a right angle straight through the straw at its end-to-end center as shown, not through the side-to-side center, but close to one side. This needs to be done very neatly.
3. The straw now has a top side, where the needle is. Remove the needle and carefully make two more holes with it, one near each end going horizontally through the center of the straw.

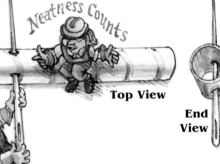

> "When you get to the end of your rope, tie a knot and hang on."
> —Franklin D. Roosevelt

4. Cut four lengths of thread 12 inches (30 cm) long.
5. Take two lengths of the thread, put them together, and tie a knot about 2 1/2 inches (6.5 cm) from one end.

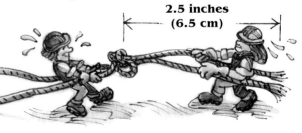

Create Your Own / Balance for Small Weights

6. Thread the needle with the unknotted end of the double thread, and put it through the holes at one end of the straw. Remove the needle, leaving the thread in the holes. Tie a second knot in the double thread, like the first, 2 1/2 inches (6.5 cm) from the free end.
7. Repeat steps 5 and 6 with the other pair of threads at the other end of the straw.
8. Put the straight pin into the original, centered hole in the straw and tape the sharp end of the pin snugly to the end of the ruler. The straw can rotate around the pin.
9. Place the ruler on a table or countertop so that the end with the pin sticks out off the edge. Tape the ruler to the table. If possible, choose a location where you can leave your balance set up for a while.

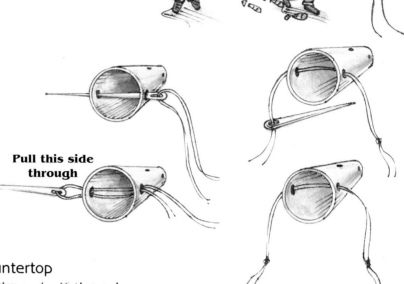

Pull this side through

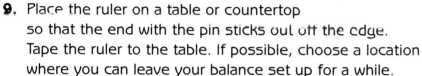

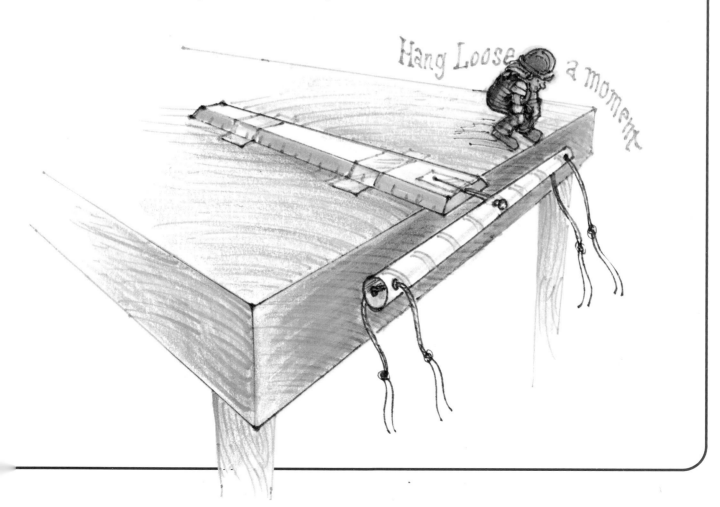

10. Separate the free ends of one of the double threads to tape them to the muffin paper with small pieces of tape. Curve the thread where it is taped so it won't easily slide out of place.
11. Repeat step 10 with the other double thread and the second muffin paper.
12. Probably, your balance is not yet balancing. It is very sensitive to any difference between the two sides. So now you need to adjust it by adding small bits of tape to the outside bottom of the basket on the *high* side. Add a bit of tape, patiently let the balance come to rest, then see if you need to add more.
Notice that just breathing on it will cause it to move.

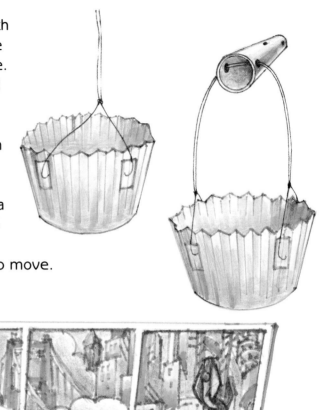

II. Using the Balance

Once you get the balance level it is ready for use. So that the baskets don't spin around, support them with your hand as you add objects. To weigh an object, put it in one of the baskets. Gradually add the weights—mung beans, lentils, or beads—to the other side until the two sides balance. This is a light-duty tool, so don't use objects heavy enough to bend the straw.

Although it is cheap and easy to make, this little balance is *amazingly* reliable. See what happens when you put just one seed on one side. For good results, be sure that the straw does not touch the end of the ruler.

If you want to know weights in a standard unit such as ounces or grams, take your beads (or whatever you are using) to school or somewhere that has a scale for small weights. Find out how much 100 beads weigh. Grams are much easier to use than ounces. Let's say you discover that 100 beads weigh* 42 grams. Divide 42 by 100 to find out that each bead weighs 0.42 grams. Now you can find the weight in grams of any beadweight. Just multiply the number of beads by this conversion factor. If you find an object weighs 148 beads, its weight in grams is

$$148 \text{ beads} \times 0.42 \text{ grams per bead} = 62.2 \text{ grams}$$

or

$$\text{number of beads} \times \text{standard weight per bead} = \text{standard weight}$$

---

* Grams are actually units of mass, not weight. The difference between mass and weight is so scrambled up in common usage that it is a challenge to sort out. Fortunately, as long as you don't go traveling to some other planet where gravity is different, it won't cause you trouble. To learn more about mass vs. weight and the role played by gravity, go to page 210 in the appendix.

Vocabulary and concepts appear on page 186.

17. Water in Ketchup?

Jerry: My parents are sending me to camp.
Larry: Why? Do you need a vacation?
Jerry: No, I don't, but they do!

Would you pour water on your French fries, hamburger, or onion rings? Yuck! Probably you wouldn't, but ketchup does have water in it. What is your guess . . . uh, I mean estimate, of how much water ketchup contains? Is ketchup 2% water? 20%? 50%?

In this investigation you will find out just how much water ketchup has and, if you like, you can compare the water content of different brands.*

*This idea came from Francine Pivinski of North Potomac, Maryland

114

Inspections and Dissections / Water in Ketchup?

Materials

New bottles of one or more brands of ketchup (squeeze bottles, if possible)

Several foil baking cups (for cupcakes and muffins), one more than the number of samples you decide to test

Scale or balance sensitive to 0.1 g or the balance from "Balance for Small Weights" on page 109 and a supply of mung bean seeds, lentils, or beads (they will be called "seeds")

1 permanent marker
1 cookie sheet
Your oven

Note: Mung beans are also used in "Spunky Sprouts" and "Seed Survival" They are available at health food stores and some grocery stores.

"A wise man never knows all. Only fools know everything."
—African Proverb

I. Water in Ketchup? Experiment

Activity time once balance is made: 35 minutes at first, depending on how many samples you are doing, and then 20 minutes 8 hours later

The big picture is that you are going to weigh ketchup samples, dry them in the oven, then weigh them again and calculate the percentage of water in each. You can do this for one brand or as many brands as you like. You might want to compare two or three of the leading brands. Careful measuring is all-important in this investigation!

1. Figure out what brands and how many samples of ketchup you are going to test and write a clear statement of the purpose of your investigation on the record sheet. Do more than one sample for each brand. The more samples you test, the closer you will get to the true water percentage when you average your results. Though the data table just has room for six samples, you can use extra paper to extend the table.
2. Write an estimate of how much water you think ketchup has and, if doing more than one brand, how your brands will compare. Tell your reasons why.
3. Set up your balance in the kitchen or other area where the flooring is washable. Hey, no one's saying you're clumsy. *Just* in case . . .
4. Use the permanent marker to write a sample number or code such as "HZ1" for "Heinz sample #1" on the top inside of each foil.
5. Holding the cap shut, shake the ketchup bottle well. This is important! You can't test a whole bottle of ketchup, so you want your sample of ketchup to represent the whole bottle.

6. *If you are using a commercial scale or balance,* zero the scale with the first foil baking cup on it before doing the weighing. (This will simplify the math later. You won't have to subtract the weight of the cup because it never was included in the weight.) Put a rounded teaspoonful of ketchup into the baking foil on the weighing pan and find its weight. Record your data on the record sheet. Repeat with all the samples. Try to use the same amount of ketchup for each sample, but never assume that the weights are the same. Weigh them all. Don't rezero your scale until you are completely done with the experiment. Skip to step 12.

   If you are using the homemade balance, continue through the following steps.

7. Put a rounded teaspoon of ketchup into a labeled foil baking cup. Put some seeds into a second foil cup.

8. Make sure your balance is level, then put the ketchuped foil cup into one of the balance's baskets, while supporting that side with your hand.

9. Having three hands for this is helpful. Got a friend who can loan you one? Place the foil cup with the seeds on the other side and adjust the number of seeds until the balance levels out. Remove both foil cups from the balance.

10. Count the number of seeds in the cup and record it on the record sheet. Note that often the balance won't level out perfectly and you need to estimate what would do it. If 61 seeds is too little and 62 seeds is too much, record 61.5. If it is closer to 62 than 61, write 61.8 or 61.9 seeds. Your units of weight for this experiment are seed-weights or "seeds."

11. Repeat steps 7–10 for each sample you are testing. The reliability of your results is better the more samples you have.

12. Place the ketchup samples on a cookie sheet. Wiggle each foil cup back and forth to get the ketchup to spread out a bit. Don't touch the ketchup though, since it is already weighed. If you lose any, your data won't be right.

> *"Denial ain't just a river in Egypt."*
> —*Mark Twain*

13. Place the cookie sheet with ketchup samples into the oven set at the lowest setting, which is probably between 140°F (60°C) and 170°F (77°C). Let the samples stay in the oven for 8 hours. This will drive out the water by increasing the rate of evaporation. The result will look like fruit leather.
14. When the ketchup has dried for 8 hours or more, remove the cookie sheet and samples from the oven and allow samples to cool for about 10 minutes. Weigh the samples the same way you did before and record your data.
15. To calculate the percent of water, first subtract the final weight from the original to get the weight of water lost. Then divide the water weight by the original weight and multiply by 100. See the example in the data table.
16. Make a circle graph to show the percentage of water in one brand of ketchup. Make a bar graph to compare the water percentages of several brands. For help in making graphs, see the Graphing Appendix on page 204.
17. Now think about any sources of error you might have had, that is, anything that may have kept your results from being perfect. Did all the samples get spread out so that they were even in thickness? Did the oven seem to dry all areas evenly? List your sources of error on the record sheet.
18. Write a conclusion on the water content of ketchup based on your percents and your awareness of the sources of error.

II. Other Investigation Ideas

Activity time: the same as for Water in Ketchup? experiment

You can find the water content of many foods and beverages the same way. Apples, milk, popcorn, cookies, lettuce, cucumber, salami, cheese, bread, egg white, and egg yolk are just a few of the possibilities. If it's a solid food, slice it very thin. Evaporation only occurs at the surface of a substance, so you want to have as much surface area touching the air as possible.

Vocabulary and concepts appear on page 187.

Water in Ketchup? Experiment

Purpose:
The purpose of this investigation is to find out _____
_____

Water Content Estimate:
_____
_____

Data and Calculations:

	EXAMPLE	
Brand or sample #:	HZ1	_____ _____ _____ _____ _____ _____
Original weight of ketchup:	72	_____ _____ _____ _____ _____ _____
− Weight of dried ketchup:	− 49	_____ _____ _____ _____ _____ _____
= Weight of water evaporated:	23	_____ _____ _____ _____ _____ _____
(Units of weight: _____)		

	(WEIGHT OF WATER	÷	ORIGINAL WEIGHT KETCHUP)	× 100 =	PERCENT WATER
Example:	HZ1 (23	÷	72 = 0.31944)	× 100 =	32%
Brand/Sample # : _____	(_____	÷	_____)	× 100 =	_____%
Brand/Sample # : _____	(_____	÷	_____)	× 100 =	_____%

Water in Ketchup? / Record Sheet

	$\begin{pmatrix} \text{Weight} \\ \text{of Water} \end{pmatrix}$	÷	$\begin{pmatrix} \text{Original Weight} \\ \text{Ketchup} \end{pmatrix}$	× 100	=	Percent Water
Brand/Sample # : _____	(_____	÷	_____)	× 100	=	_____ %
Brand/Sample # : _____	(_____	÷	_____)	× 100	=	_____ %
Brand/Sample # : _____	(_____	÷	_____)	× 100	=	_____ %
Brand/Sample # : _____	(_____	÷	_____)	× 100	=	_____ %

Observations/Notes:

Sources of Error:

_____
_____
_____

Conclusion:

_____
_____
_____
_____

18. Seed Survival

What do navy beans, peach pits, dandelion fuzz, coconuts, and hitchhiking burrs all have in common? They are all seeds.* They are wildly different in size, texture, and method of getting around, but all have the ability to grow into a new plant. Seeds contain baby plants and food for the plants' first days, so they are also prized food for many animals including people. Do this activity to find out how many of your foods are actually seeds, what's inside a seed, and how certain treatments of seeds affect sprouting. Then quit bothering them, plant some interesting ones, and enjoy watching them grow.

*The white part of dandelion fuzz and the surface of most burrs are not actually part of the seeds.

Inspections and Dissections / Seed Survival

Materials

A few peanuts (in the shell, if possible)
A few large beans such as fava beans or lima beans (fresh, canned, or frozen)
Hand lens, optional
320 dry beans such as lentils or mung beans found in health food sections or health food stores and also used in "Sprout Jungle" on page 89 and "Water in Ketchup" on page 114.
1 small to medium size cooking pan
1 metal strainer that fits inside the pan, *or* a tea ball, *or* 6 tea bags

Microwave oven
Stop watch or timer
16 paper cups
1 permanent marker or crayon
Paper towels
2 avocados
6 toothpicks
Several pots and potting soil
Seeds of your choice to plant
2 drinking glasses

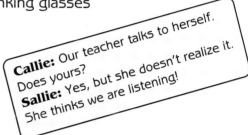

Callie: Our teacher talks to herself. Does yours?
Sallie: Yes, but she doesn't realize it. She thinks we are listening!

I. Seed Awareness

Activity time: 15 minutes

See how many foods you can find in your kitchen that are seeds or are made mostly from seeds. You might be surprised. Grains, beans, and nuts are all seeds. Many spices and oils are made from seeds. Circle the ones you find on the record sheet.

II. Seed Dissections

Activity time: 10–15 minutes

Inspect the following (and other) seeds to find the *seed coat*, the *embryo* (the baby plant), and the food supply. Sometimes you can even see the leaves and the root of the embryo.

1. **Peanuts.** Most of the peanuts you find in the grocery store are roasted. That's fine for this section, but you won't be able to raise plants from them. The peanut shell, with its contents, is the ripened ovary of the flower. The dark brown papery stuff surrounding each edible peanut is the seed coat. When you pull a peanut into two parts, one half has a groove at one end, and the other half has the embryo. It looks like a tiny bearded face or a fish. The "tail" of the fish or the "beard" of the old man is formed by the embryonic leaves. Look at the embryo through a hand lens, if you have one, to see these better. The rest of the nut is food for the plant-to-be.
2. Sketch the peanut, labeling its parts on the record sheet.

3. **Lima or fava bean.** On the inner curve of a bean, you can see a scar where the seed was once attached by a little stalk to its pod. The seed coat is fairly thick but peels off easily from the rest of the seed. The embryo is easy to find inside when you split the seed in half. See if you can distinguish embryo root from the embryo leaves. The rest of the seed is food supply.
4. Sketch and label the parts of the bean.

II. Seed Survival Experiment

Activity time: 1 hour to do all treatments and 10 minutes a day for a few days after that

How hardy are seeds? Can they sprout after being frozen or boiled? Can they handle microwave radiation? You will be boiling, freezing, and microwaving samples of dry seeds, then allowing them to sprout so you can compare survival rates in the different conditions.

1. Figure out your statement of purpose and write it on the record sheet.
2. State a prediction of the outcome on the record sheet.
3. Choose dry, uncooked seeds such as mung beans or lentils. There are many seed possibilities, but don't use canned or frozen seeds for sprouting.
4. Count out 20 seeds into each of 16 paper cups. Throw away any seeds that are a strange color or shape. Write the planned treatment on the side of the cup with a permanent marker. A series of treatments are suggested below but you might prefer to work out your own plan.
5. One cupful of seeds gets no treatment except to sit on the countertop. This is your *control* group. Label the cup "C."
6. Three cups will go into the freezer. Label them "F5," "F30," and "F60" for 5, 30, and 60 *minutes*. Put them in the freezer and watch the time or set a timer to take them out at the proper time.
7. Boil some water in the pan. The seeds from six cups will be going into boiling water for 4, 8, 16, 32, 64, and 128 *seconds* so label six cups "B4," "B8," and so on. To get the seeds into and out of the boiling water all at once, put the seeds, one bunch at a time, in a metal strainer or in a tea ball. Start the stopwatch as you lower them into the water,

Warning:
Be careful! Boiling water can cause burns.

and then remove at the proper time. Another way to do this is to take apart a tea bag, remove the tea and re-staple the bag with the seeds inside. How clever! Return the seeds to their labeled cups.

8. The remaining cups will be microwaved on high power as follows: 4, 8, 16, 32, 64 (1:04), and 128 (2:08) *seconds*. Label them "M4," "M8," "M16," and so on. This part is easy. Just put one cup at a time into the microwave, set for the appropriate time, and zap those poor seeds.

9. After the treatments, add water to each of the 16 cups (fill cups halfway) to soak the seeds overnight or for 8 hours to 12 hours.

10. After the soaking, drain the extra water and rinse the seeds.

11. Now you will let the seeds sprout in their paper cups. Stuff some crumpled paper toweling into each cup on top of the seeds. Add just enough water to wet the toweling. Each day for several days, remove the wad of toweling, rinse the seeds, and then replace the toweling, keeping it moist. Watch for signs of sprouting.

12. Three days after the first seeds sprout, declare that the time is up and count the sprouted seeds. Write the count on each cup. Then transfer the data to the record sheet. Calculate the percent germination for each treatment. Divide the number germinated by the number of seeds in the cup and multiply the answer by 100. For example, if 17 of 20 seeds germinated,

17 ÷ 20 = 0.85 and 0.85 × 100 = 85 or 85%

13. List any sources of error on the record sheet. What flaws did you have in the design of your experiment that could lead to imperfect results? Did all the samples stay equally moist during sprouting?
14. Create a bar graph with 16 bars showing the germination percents for each treatment. You can also make line graphs of the microwave and boiling data. For these the horizontal (x) axis would be labeled "Time Microwaved" or "Time Boiled" and the vertical (y) axis would be labeled "Percent Germinated." See the Graphing Appendix on page 204 for help in making bar and line graphs.
15. What have you learned about seed survival after freezing, boiling, and microwaving? Write your conclusion on the record sheet.

III. Other Investigations

There are other seed germination variables you could investigate. Pick one and design your own experiment. In the preceding experiment, we investigated three different treatments, but each group of seeds differed from the control group by just one variable. We didn't, for example, try both microwaving and soaking in vinegar with one group and then freezing and adding salt with another group. You change just one variable at a time so you can tell what treatment was responsible for the sprouting results.

- What if the preceding treatments were done *after* soaking instead of before? Are seeds more resistant at one time compared to the other?
- How does the depth at which seeds are planted in soil affect germination percent?
- Will seeds germinate if the seed coat is removed after soaking?
- Will seeds germinate if part of the food supply is removed after soaking?
- How strong are germinating seeds? What is the effect of putting weights such as pennies, quarters, or something heavier on the soil above planted seeds?

Keep track of your results on a record sheet of your own making. Write your conclusion.

IV. Avocado Growing

You *gotta* try this! Avocados have a huge seed that is amazing to watch grow. You can't be in a hurry, though. It takes about 4 weeks to sprout. But once it starts, the growth is rapid, beginning with a straight spike that grows about a foot tall before leafing out.
1. Buy a couple of avocados at the grocery store. An avocado is a fruit.
2. When they ripen, eat the fleshy part and save the pits. The pit is the seed.
3. Wash the seeds and gently pick off the brown seed coat.

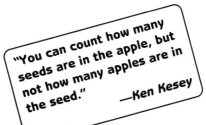

"You can count how many seeds are in the apple, but not how many apples are in the seed."
—Ken Kesey

Inspections and Dissections / Seed Survival

4. For the next 3 to 5 weeks the seeds need to sit in water. Make a support system for each seed from three toothpicks. Poke the toothpicks part way into the side of the seed so that the seed can sit pointed end up on a clear glass of water. It should be kept about half in and half out of the water.
5. Keep the glasses with water out of direct sunlight in a spot you will notice every day. Add water as needed and change the water about once a week.
6. So much time will pass without change that you will wonder if the seed will ever sprout. Keep supplying water! The first change you will notice is the growth of roots at the bottom. Once this happens, more is about to come!
7. When the roots are about 3 inches long, it is graduation time for the avocado. Move the plant to a larger container such as a flower pot with real soil under and around the bottom half of the seed. Be gentle arranging soil around the roots and leave the top half of the seed above the soil surface. Leave the toothpicks in place.
8. Keep the soil damp but not wet and put the plant in a sunny spot. Be inspired by the confident growth of this remarkable plant.

V. Growing Other Seeds

Sometimes we forget that most seeds in our foods and in our yards can grow if given the proper conditions. What do you want to try? Maple seeds? Acorns? Grapefruit or lemon seeds? Pumpkin seeds? Peanuts (raw, not roasted)? Mustard or pepper seeds? Find a milk carton, butter tub, or coffee can; fill it with potting soil; plant the seeds under about 1/2 inch (1.3 cm) of soil; and keep the soil moist. Write down the date planted and then record when you first see a green shoot. (See "Sprout Jungle" on page 89.) Try keeping a sketched record of the plant growth. Make a drawing every day at the beginning of the growth of the green shoot and then once a week after you have large leaves.

Vocabulary and concepts appear on page 188.

Observations, Ideas, Sketches, Questions

Circle any of the following seeds or seed products you have in your kitchen. Write in others you discover:

Green peas, corn, popcorn, peanuts, peanut butter, almonds, almond extract, cashews, pecans, walnuts, brazil nuts, chestnuts, macadamia nuts, pumpkin seeds, watermelon seeds, sunflower seeds, poppy seeds, sesame seeds, sesame paste, whole black pepper, ground black pepper, red pepper, lentils, spilt peas, white rice, brown rice, wild rice, wheat flour, rye flour, cornmeal, corn starch, oatmeal, cream of wheat, corn flakes, wheat flakes, granola, barley, cocoa, coffee, expresso, mustard seed, dill seed, coriander, nutmeg, cumin, celery seed, caraway seed, corn oil, peanut oil, safflower oil, canola oil, sunflower oil, coconut oil, coconut, lima beans, fava beans, garbanzo beans, black eyed peas, navy beans, red beans, black beans, mung beans, alfalfa seeds, soybeans, soy milk, tofu, wheat bread, white bread, rye bread, crackers, noodles, bagels, muffins, cracked wheat

Other: _____

Number of seed products in your kitchen: _____

"Every great advance in science has issued from a new audacity of imagination."
—John Dewey

Sketches of Seeds:
It's easier to draw and see the seeds if you make the sketches bigger than the real thing.

Seed Survival Experiment

Purpose:
The purpose of this experiment is to find out _____
_____

Prediction:
_____

Data:

Treatment	Observations	Number of Seeds Germinated	Percent of Seeds Germinated

Sources of Error:
_____
_____

Conclusion:
_____
_____

19. The Case of the Telltale Fingerprint

Would you recognize your own hands by their fingerprints? Could you solve a crime using fingerprint evidence? First, of course, you would need to have a fingerprint record to compare to. In this activity, you will prepare good sets of fingerprints, analyze them, learn to develop natural prints, and use them to identify a thief. The thief might be you!

Materials

White paper
Several file cards
1 wooden pencil
Wide (3/4 inch or 2 cm) clear tape
1 hand lens, helpful but not required
1 piece of black paper
Baking soda
1 spoon
1 deck of cards
1 drinking glass or jar, clean and dry
1 bandanna or large napkin
Timer

Q: Why didn't the skeleton go to the dance?
A: He had no body to go with.

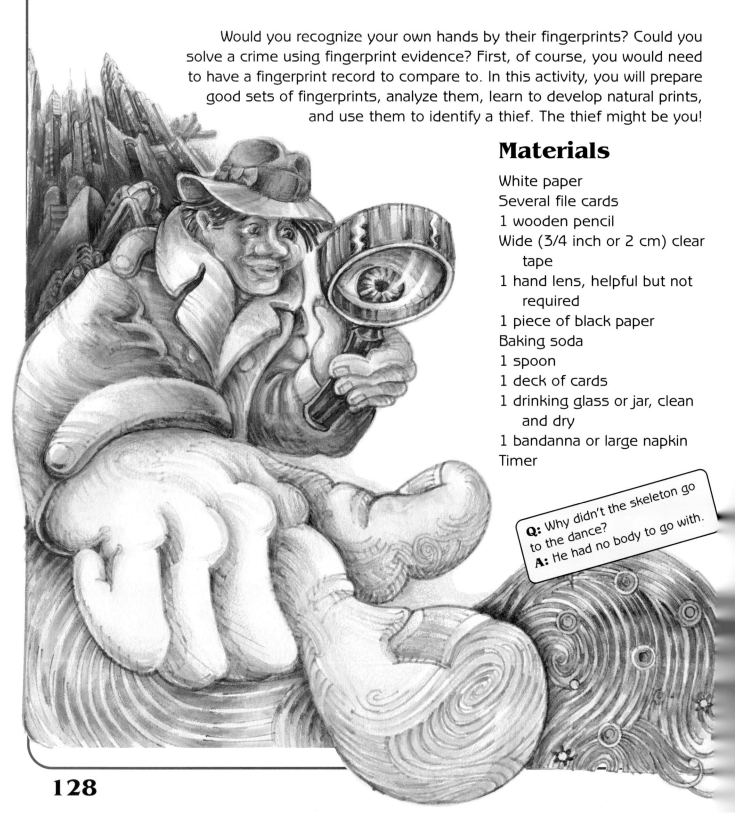

I. Prepare Your Own Fingerprints

Activity time: about 20 minutes

1. Hold your pencil as you would a knife and rub the sharpened tip back and forth on the file card (or a piece of paper) to completely darken an area about 2 inches by 2 inches (5 cm by 5 cm).
2. Tear off fifteen 1.5 inch (4 cm) lengths of clear tape and stick each of them by a corner to the edge of your table so they are handy.
3. Rub the pad of one of your little fingers around in the black spot until it is well coated with pencil "lead." (It's really the mineral graphite.)
4. Apply the sticky side of one piece of tape to the end of your finger as shown. Press the tape down so that it covers as much of your fingertip as possible.
5. Remove the tape and put it sticky-side-down on some scratch paper. You want to make clear, not-too-dark, not-too-light prints. Make practice prints this way with different fingers until you get three good quality prints in a row.
6. Now wash your hands for a fresh start to create your reference set on the record sheet. Start again with your little finger. Before you tape down a print, check to see if it's neat by holding it in front of something light-colored. If the print is smudged, throw that one away and try again. If it looks neat, stick the tape onto the record sheet in the proper spot, fingertip end up.

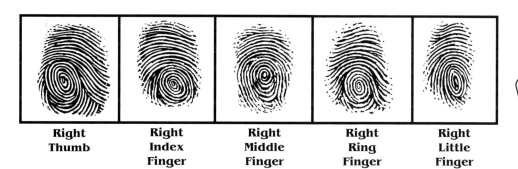

| Right Thumb | Right Index Finger | Right Middle Finger | Right Ring Finger | Right Little Finger |

"It's kind of fun to do the impossible." —Walt Disney

7. Repeat this process with ring, middle, index finger, and thumb. Once you finish with one hand, wash your hands again, then continue with your other hand, starting with the little finger.
8. Find a few other people willing to be fingerprinted. Help them create their own fingerprint sets to add to your print file. Use large file cards or plain paper for more sets once you have filled the record sheet.

II. Analyzing Prints

Activity time: about 15 minutes

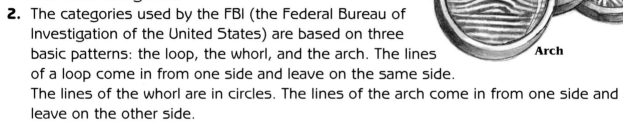

1. Now you have at least three sets of fingerprints or 30 individual prints to examine. It is helpful to look at them closely with a hand lens. While no two fingers have the same print, even from the same person, their patterns can be grouped into a few categories.
2. The categories used by the FBI (the Federal Bureau of Investigation of the United States) are based on three basic patterns: the loop, the whorl, and the arch. The lines of a loop come in from one side and leave on the same side. The lines of the whorl are in circles. The lines of the arch come in from one side and leave on the other side.
3. Use these categories or, if you prefer, create your own categories for the prints you have. Come up with names and descriptions for each category.
4. Note the category of each print on your record sheet. Now you can quickly tell what category a new print belongs to. This speeds up the process of print identification.

III. Developing Latent Prints

Activity time: 15 minutes

Natural fingerprints left on objects, called latent prints, can be enhanced with baking soda and compared with the reference sets to figure out who left them. Practice this skill as follows.

1. Set up a work station on a sheet of paper. Collect a sparkling clean glass or jar, baking soda, and a spoon. Be smart about where you touch the glass.
2. If you have just washed your hands, rub your fingers over your face or through your hair to pick up natural oil. Grasp the glass as it sits on the paper, making sure to use the pads of your fingers, not just the tips. Do not move your fingers once they touch the glass so that your prints don't smudge. Release the glass and look for your latent prints by holding the glass up to the light.

"I think the one lesson I have learned is that there is no substitute for paying attention." —Diane Sawyer

3. Use a spoon to sprinkle baking soda over the glass where you think the prints are. Tap the glass on the counter a few times to get excess baking soda off. The fine powder sticks to the oils left on the glass by the skin ridges. Roll up a piece of black paper to put inside the glass. The white baking soda shows up better on the black background.
4. Compare the prints on the glass with your reference set to see if you can recognize them. Other than color, what do you notice that is different?

IV. Crime Scene

Activity time: 20 to 60 minutes

If a real crime occurs, physical evidence is collected from the scene to figure out who the culprit was. Physical evidence includes fingerprints, blood, hair and fiber samples, handwriting, voice prints, and other clues related to physical things.

You can play a forensic science game using the latent print-developing technique you just practiced. You need three to five players. The idea is to secretly assign one person as thief, have the thief leave prints on an object involved in a crime, develop the prints, and then use the prints to reveal the thief.

1. Collect a group of people to play the game. Have everyone make reference sets of their prints. Teach your friends your classification system.
2. Make up a story about the theft of a precious rare stamp. In the next room, place a clean glass upside down over a stamp (put in a normal stamp, just pretend it's worth millions). Use a bandanna or other cloth to handle the glass so you don't leave prints on it. Cover the glass with the cloth.
3. Count out as many cards from a deck as you have people and choose one of those cards, such as the ten of diamonds, to mean "thief."
4. Mix up the cards and pass them out so each player gets one. Players should keep a "poker face" (no expression—you don't want to let anyone know if you are the thief or not) when checking the card.
5. Now, one at a time, each person goes to the covered treasure. The real thief will uncover the glass, apply his or her fingers to the glass (carefully), and steal the stamp. Others will just hang out a moment. No one but the thief is to lift the handkerchief or touch the glass. (If you think the temptation for non-thieves to see if the stamp is still there will be too great, make it an *imaginary* stamp.)

6. When everyone has had a turn to go to the stamp case, remove the cover. The stamp has disappeared! Help! Thief! Call the police!
7. Now the prints can be dusted. Apply the baking soda as you did before.
8. Use the timer so that each player gets one minute to examine the glass and the sets of reference prints. Keep poker faces again as you make identifications.
9. After the evidence has been studied, each person gets a chance to accuse someone of being the thief. The thief should not reveal himself or herself until every member of the group (including the thief) accuses someone. When all the accusations are made, he or she finally makes a dramatic confession. "The butler made me do it!"
10. If your accusation turns out to be right, you win ten points. If it was wrong, you lose ten points. Pass if you don't know. (Clueless?) The thief gets ten points as long as the prints were clear enough for someone to identify him. The thief loses ten points for smeared prints that no one can identify. (Not *exactly* the way real criminals would do it.)
11. Wash the glass and do another round!

Vocabulary and concepts appear on page 189.

The Case of the Telltale Fingerprint / Record Sheet

Observations, Ideas, Sketches, Questions

Fingerprint Records

Categories:

_____
(Your name)

Right Hand: thumb   index   middle   ring   little

Left Hand: thumb   index   middle   ring   little

_____
(Name)

Right Hand: thumb   index   middle   ring   little

Left Hand: thumb   index   middle   ring   little

134

The Case of the Telltale Fingerprint / Record Sheet

_____
(Name)

Right Hand: thumb   index   middle   ring   little

Left Hand: thumb   index   middle   ring   little

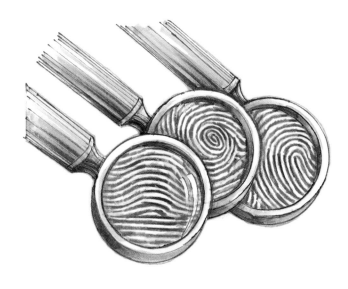

20. The Incredible Egg

Sandy: What are quadruplets?
Mandy: Four crying out loud!

An egg is a compact bundle of nutrients wrapped in a mineral coating that lets air in and out yet is amazingly strong. Have you ever studied one? In this activity, you will look through an egg without cracking it, then take a closer look at the structures inside. You will finish up by experimenting with shells and with whole raw eggs without shells!

By the way, can you imagine being the size of a chicken and laying an egg? Oooh!

Materials

Cardboard, at least 6 inches by 6 inches (15 cm by 15 cm)
Sharp scissors
10 to 12 fresh eggs (only a few of them will get used up)
1 bright flashlight
1 plate
1 spoon
1 clear drinking glass
2 coffee mugs
Paper towels
White vinegar, about 2 cups (480 mL)
Corn syrup, optional

I. Seeing Through the Shell

Activity time: about 15 minutes

When eggs are sized and graded for sale, they are inspected using a bright light. The process is called candling, since it was first done using candles. To see inside a whole egg, make a platform for the egg, then shine a bright light through it. Here's how:

1. Trace your egg lying on its side on a sheet of cardboard, holding your pencil straight up and down. Draw a second outline within the first so that the outline is the shape of your egg but a little smaller.
2. Cut on the inner line with scissors and remove the center of the oval. Your egg should be able to lie over the hole without falling through.
3. Take the cardboard, the egg, and the flashlight into a windowless room such as a closet. Place the egg on the cardboard over the opening.
4. With the lights off in the room, shine the flashlight up through the egg from below. Cool, eh? The brighter the light, the better it is.
5. Look closely. Turn the egg to see all sides. Look for distinctive shell markings that would help you identify this egg in a crowd. Can you see the yolk? The air space? Record your observations on the record sheet.
6. Just for fun, try to find your egg after it has been put back with the rest of the eggs. Have someone else mix up the eggs so you don't know which it is. See if you can identify it with the help of candling.

II. Inside the Egg

Activity time: 15–20 minutes

1. Crack the egg on the edge of a plate, then carefully pour the innards onto the plate as you pull the two halves apart. You don't want to break the yolk. Save the shell for Part III.

Inspections and Dissections / The Incredible Egg

2. Take the time to look closely and, on the record sheet, sketch what you see. After you've made your sketch, turn to page 190 for the names of the different parts. Add labels to your sketch.
3. Every living thing needs food, water, air, and protection. How would a chick embryo growing in an egg get each of these requirements?
4. Press the yolk lightly with the bottom of a spoon. Notice that, like a water balloon, it springs back to shape when you lift the spoon. It is surrounded by a membrane.
5. Press down harder on the yolk membrane until it pops.
6. Now examine the membranes on the shell. Find the air cell attached to the shell and feel it.
7. Peel the membrane away from the shell. Break the shell around the edges to pull the membrane away from the rest of it. As you look at the torn pieces of the membrane, you will see places that are half the thickness of the rest. That's because there are two membranes next to each other. Usually you peel away both together but sometimes one pulls away from the other. They are separate at the air cell.
8. Fill a clear glass with water and put some membrane pieces into the water. Poke them to remove air bubbles. Do they float or sink?
9. Put some bits of bare shell into the water. Do they float or sink?
10. Use a spoon to scoop some of the egg white to drizzle it into the water. Sink or float?
11. Do the same with some of the yolk. See how cool that looks? Does it sink or float? Record your observations on the record sheet.
12. Wash your hands with soap after handling raw eggs because they occasionally contain harmful bacteria.

What kind of soup do chickens eat when they don't feel well?

III. Mini Eggshell Lava Lamp

Activity time: about 10 minutes

1. Put white vinegar into a clear glass so that it is about 4 inches (10 cm) deep. Other kinds of vinegar will also work but they are harder to see through.
2. Remove *all* the membrane from some of the shell you have left and break the shell into about 15 pieces of different sizes.
3. Drop the shell pieces into the vinegar. Stir once.
4. Watch for at least 2 minutes and record your observations on the record sheet.
5. What's your explanation for what is happening? If you look closely, you can see what causes it. Write your hypothesis of the cause on the record sheet.
6. The action may stop after a while but can be restarted with stirring.
7. Save this vinegar for reuse in Part IV.

IV. Remove the Shell Without Breaking the Egg

Activity time: 5 minutes at first, then 10 more after 24 hours of soaking

1. Select two fresh eggs that are about the same size and shape. You will chemically remove the shell from one egg and keep the other egg handy for comparison. Set each egg into a cup. Add vinegar to *one* of the cups so that the egg is completely covered. Observe the egg in the vinegar for at least 1 minute and record your observations on the record sheet.
2. Allow the egg to sit in the vinegar for 24 hours.
3. Use a spoon to remove the egg from the vinegar. When all the shell is gone, the surface is partly clear, translucent, and golden in color. If there are still white or brown areas (depending on the original color of the shell) you can soak the egg a few hours longer in fresh vinegar. When it is ready, dab it dry with a paper towel.

"A hen is just an egg's way of making another egg."
—Samuel Butler

4. Inspect your shell-less egg gently. How does it feel? Compare it to the very lonely egg that has been sitting in the dry cup. How do the sizes compare? That membrane is pretty tough, isn't it? See what happens to the yolk when you hold the egg upright, then turn it upside down. Record your observations on the record sheet.
5. Candle the egg as you did in Part I. What do you observe?

V. Investigate the Membrane Barrier

Activity time: a few minutes a day for 9 days

1. Put the shell-less egg into a cup that is two-thirds full of corn syrup and allow it to soak overnight.
2. Remove the egg, rinse briefly, and compare it to its partner egg. Record your observations on the record sheet.
3. Now put the shell-less egg back into a cup of fresh water and let it sit overnight.
4. Remove the egg, compare to the other, and record your observations.
5. Now let the eggs sit uncovered in the refrigerator for a week on a paper towel. Compare the shell-less one with its partner-in-the-shell each day. What's going on?

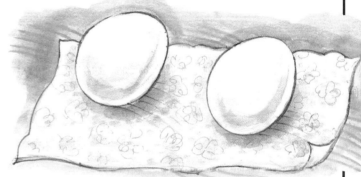

VI. Other Investigations

- Hard boil an egg (cook it in boiling water for 11 minutes, then run it under cold water) and look for all the parts you found and labeled in Part II.
- Hard boil shell-less and regular eggs and see how they compare inside.
- Compare the spinning of a hard-boiled egg and a regular egg (both with shells). What could explain the difference?
- If you have access to a balance, compare the weight loss that occurs in eggs with shells compared to eggs without shells. What effect, if any, does the eggshell have on water loss by evaporation? Record the weights of regular and shell-less eggs every day for a week or more. What do you find? Your data would make a nice line graph with a line for each egg. See the Graphing Appendix on page 204 for information on making a line graph.

Vocabulary and concepts appear on page 190.

Observations, Ideas, Sketches, Questions

I. Candling Observations:

II. Egg-on-a-Plate Sketch:

YOUR HYPOTHESIS OF HOW THE CHICK GETS THIS ITEM

Food	
Water	
Air	
Protection	

Egg Part	Float	Sink
Membrane		
Shell		
White		
Yolk		

The Incredible Egg / Record Sheet

III. Lava Lamp Observations:

Lava Lamp Hypothesis:

IV. Observations of Vinegar-Soaked Eggs:

The first minute:

After 24 hours:

V. Shell-less Soaking and Drying Observations:

VI. Notes on Other Investigations:

21. Horoscope Wisdom

Leo (Born July 23–August 23):
Today is a good day to take care of money matters. Family tensions will begin to ease if you act from the heart. Your current project is about to take a turn for the better but beware of letting success go to your head.

Some people believe in horoscopes, while others think horoscopes are baloney. A lot of people aren't sure either way. But does anyone test them? Is testing them even possible? Here is an experiment that will give you some insight on the wisdom of horoscopes.

Materials

Horoscope from a daily newspaper, in print or from the Internet
Computer and printer OR scissors, tape, and the use of a copy machine

Horoscope Experiment

Activity time: 2–4 hours

The big picture is that you will ask people, your "subjects," to rate how well two versions of yesterday's horoscope fits them. One of the two versions will be a horoscope straight from the newspaper. The other version will look just as real but will have all the messages listed under the wrong zodiac signs. The average rating for "fit" will be found for the proper and the scrambled horoscopes. If horoscopes fit because they really match the people born under that zodiac sign, then the average rating for the regular horoscopes should be significantly higher than the average rating for the scrambled horoscope.

1. Make a prediction about the outcome of the experiment. Record it on the answer sheet.
2. Figure out how you will find lots of subjects to take your survey. Is there a club meeting, party, or other event coming up soon? Between 6 and 15 subjects is fine. But the more subjects you have, the more reliable your results will be. Can you poll students and teachers in the cafeteria? Could a classroom teacher help you by having students take them in class?
3. Get a horoscope from a newspaper the day *before* the day you are going to collect your data. The horoscopes are usually found in the section with the TV listings, comics, and advice columns. On the Internet some of the horoscopes are long. You want one where the messages for the day for every sign will fit on one page.
4. Copy the horoscope, and then use the copy you have made to create a second version. Cut and tape (or copy and paste on the computer) the messages for each sign of the zodiac so that each one is under a new, incorrect sign, two or three signs apart. Work neatly so that your cutting and taping will not show once the paper is copied. Label one horoscope "A" and the other "B." The date of the horoscope (yesterday) should show for both versions. And, since some people don't know their zodiac sign, be sure your horoscope shows the birth dates for each sign.

Tuesday, August 20, 2002
Horoscope A

Aries (March 21–April 19)
Spend as much time as you can with family. They need your level-headed influence. Take care of settling money questions now. It will pay off later.

Taurus (April 20–May 20)
Love will catch you by surprise today. Don't be shy! Patience will be needed in your work life. Attend to health concerns, even though they seem minor.

Gemini (May 21–June 21)
A misunderstanding is cleared up today. Celebrate the breakthrough! Look into opportunities opening up. Your skills will fit better than you first think.

Cancer (June 22–July 22)
It's time to end a relationship that is causing you difficulty. Speak honestly but firmly and you will be respected. Your enthusiasm brings you recognition on a big project.

Leo (July 23–Aug. 22)
Your good work will be noticed by those in authority today. You can help an unhappy friend but don't let the troubles of others get you down.

Virgo (Aug. 23–Sept. 22)
Someone close to you is struggling. Lend an ear but avoid giving advice. Take action to renew an old contact. Watch your choice of foods to keep energy high.

Libra (Sept. 23–Oct. 23)
Admit your fears to someone close and you will make surprising progress in overcoming them. Splurge on something you have been wanting. Take time to organize a key space.

Tuesday, August 20, 2002
Horoscope B

Aries (March 21–April 19)
Your good work will be noticed by those in authority today. You can help an unhappy friend but don't let the troubles of others get you down.

Taurus (April 20–May 20)
Someone close to you is struggling. Lend an ear but avoid giving advice. Take action to renew an old contact. Watch your choice of foods to keep energy high.

Gemini (May 21–June 21)
Admit your fears to someone close and you will make surprising progress in overcoming them. Splurge on something you have been wanting. Take time to organize a key space.

Cancer (June 22–July 22)
Spend as much time as you can with family. They need your level-headed influence. Take care of settling money questions now. It will pay off later.

Leo (July 23–Aug. 22)
Love will catch you by surprise today. Don't be shy! Patience will be needed in your work life. Attend to health concerns, even though they seem minor.

Virgo (Aug. 23–Sept. 22)
A misunderstanding is cleared up today. Celebrate the breakthrough! Look into opportunities opening up. Your skills will fit better than you first think.

Libra (Sept. 23–Oct. 23)
It's time to end a relationship that is causing you difficulty. Speak honestly but firmly and you will be respected. Your enthusiasm brings you recognition on a big project.

5. Make a form like this one for each person to complete. You may choose to include question III or not. It will give you interesting information, but it is not necessary for the experiment.

HOROSCOPE EXPERIMENT QUESTIONNAIRE

I. Your sign:_____

II. Please read the two horoscopes for your sign from yesterday and think about how well each fit you and your life as things went yesterday. Circle one of the ratings below for each version.

Horoscope A

1	2	3	4	5
did not fit me at all		medium fit		very good fit

Horoscope B

1	2	3	4	5
did not fit me at all		medium fit		very good fit

III. Do you believe there is some truth in horoscopes? Circle one:

Yes   No   I don't know.

"A free society is a place where it is safe to be unpopular."
—Adlai Stevenson

Wyler: What's an optimist?
Tyler: A hope addict.

Psychology and Beliefs / Horoscope Wisdom

6. Make copies of the horoscope sheets and forms as needed. You must collect all your data in one day since the subjects have to think about "yesterday."
7. Tell your subjects, "I am doing a study of horoscopes. Please complete this form honestly. Thank you for your help." Each subject then reads his or her own horoscopes for the previous day and rates them. Give the subjects enough time to read each of their horoscope messages, but don't give them so much time that they read the messages for the other signs. Should they realize that the two forms have the same messages under different signs, it doesn't matter too much, as long as that is all they know. If a subject wants to discuss the design of the experiment, tell them that you would be glad to do so *after* the study is completed. Keep the forms to tally later.
8. You need to keep the testing conditions for your two versions fair. Sometimes there is an advantage to being looked at first or last. To make sure that an "order effect" does not occur, have half your subjects read version A first, and the other half read version B first. Then, if there is an advantage to being first or second, both versions will share it.
9. When the data collection is done, find the average ratings for each version. Do this by adding together all the ratings for one and dividing by the number of subjects. Do the same for the other version and then compare.
10. Also count the number of subjects who gave higher ratings to each version. Turn your counts into percents by dividing them by the total number of subjects and multiplying by 100.
11. Your data lends itself to two bar graphs, one to compare the average ratings for each horoscope, and one for the percent of subjects scoring higher on each version. See the Graphing Appendix on page 204 for help making a bar graph.
12. Think about any sources of error you may have had. What flaws did you have in the design of the experiment that could lead to imperfect results? Was there anything different about the looks or presentation of one version compared to the other? List your sources of error on the record sheet.
13. What do you conclude? Write a statement that compares the ratings and percents and explains what they tell you about horoscopes.

> **"Everyone has inside of him a piece of good news. The good news is that you don't know how great you can be! How much you can love! What you can accomplish!"**
> —*Anne Frank*

Vocabulary and concepts appear on page 192.

Horoscope Widsom Experiment

Purpose:
The purpose of this experiment is to find out how closely horoscopes match people of the correct zodiac sign. This will be done by comparing the reported fit of horoscopes for the correct and incorrect signs.

Prediction:
_____
_____

Data:

COLLECTED RATINGS FOR FIT

Subject	Normal Horoscope	Scrambled Horoscope
1		
2		
3		
4		
5		
6		
7		
8		
9		
10		
11		
12		
13		
14		
15		
16		
Total		
Average (= total/the number of subjects):		
Number of Subjects with Higher Score:		
Percent of Subjects with Higher Score (= # with higher score/total # of subjects):		

Other Observations:

Sources of Error:
_____
_____
_____
_____
_____

Conclusion:
_____
_____
_____
_____
_____

22. Color Matters

Most people like food that is attractively served. Does a nice food appearance actually improve food flavor or just make eating more pleasant? Your mouth contains taste buds, and, if you have done "Flavor Detective" on page 16, you know the role of the nose. So what about the eyes? Does vision contribute to the perception of flavor? Find out by doing taste tests of soft drinks that differ only in color.

Materials

2 bottles of a clear soft drink such as Sprite or 7-Up
Food colors
Paper ⎫
Tape ⎬ for labels
Marker ⎭
Small paper cups
The use of a copy machine or a computer with printer, if possible

Q: Who is beautiful, gray, and wears glass slippers?
A: Cinderelephant!

I. The Effect of Color on Flavor Experiment

Activity time: 1–2 hours

The big picture is that you are going to prepare two versions of a soft drink, one that has an appealing, normal color and one that has an unusual, not-so-appealing color. Subjects will rate the flavors of the two versions. To ensure that they try to separate the flavor experience from how the sample looks, they will also rate the appearance of the versions. The subjects should not know how the ingredients of the versions compare. If color has no effect on flavor, the average flavor ratings of the two versions should be very close.

1. Pretest your two colors in water to find out how to get the effect you want. Use an amount of water equal to the soft drink in one bottle. Count the drops you add. The "normal" color may be clear, if you like.

2. Color the soft drinks in their bottles the way you want them. Write down on the record sheet your formulas, how many drops of each food color you added for each drink color.

3. Peel off the labels on the bottles and replace them with your own. Label one bottle "A" and the other "B." You want your subjects to be "blind" to the identity of the drinks.

4. Chill the drink bottles in the refrigerator.

5. Figure out how to get lots of subjects to take your taste test. Is there a club meeting, party, or other event coming up soon where you could set up a little booth? Somewhere from 6 to 15 subjects would be fine. But the more subjects you have, the more reliable your results will be.

6. On the record sheet, write a clear purpose of the experiment and your prediction of what this test will find.

> "The way I see it, if you want the rainbow, you gotta put up with the rain."
> —Dolly Parton

7. Prepare a form something like this for each subject to complete:

Drink Sample	Flavor Rating (circle one)	Appearance Rating (circle one)
A	1 2 3 4 5 6 7 8 9 10 (very bad)　　　　　(very good)	1 2 3 4 5 6 7 8 9 10 (very bad)　　　　　(very good)
B	1 2 3 4 5 6 7 8 9 10 (very bad)　　　　　(very good)	1 2 3 4 5 6 7 8 9 10 (very bad)　　　　　(very good)

8. You must keep the testing conditions for your two versions fair. They should be the same temperature and presented in the same way. Sometimes a food or drink tasted first or last in a taste test has an advantage. To make sure that an "order effect" does not occur, have half your subjects taste one colored sample first and the other half taste the other one first. Then, if there is an advantage to being first or second, both versions will share it.

9. When a subject comes to your test station say something like, "I am going to give you two soft drinks to taste. On a scale of one to ten, with ten being the highest, please rate the flavor and the appearance of each drink."

10. When you have tested all your subjects, transfer the data to your data table and work out the averages.

11. To find the average ratings, add the numbers in one column and divide by the number of subjects.

12. Did people rate the flavors of the two drinks differently? How do the taste ratings compare to the appearance ratings?

13. To display your results, make a bar graph showing your average taste and appearance ratings. See the Graphing Appendix on page 204 for more information on making a graph.
14. Take a hard look at any weaknesses your study had. Did you eliminate possible order effects? Anything that kept your test from being fair should be listed as a source of error on the record sheet.
15. Write a conclusion on the effect of color on food flavor. Use your statistics to support your statement.

II. Design Your Own Experiment on Food Appearance

Activity time: will vary

Use these ideas to get your own thoughts going. Be sure to write a clear purpose for your experiment to keep yourself on target with the planning. Create a table that fits your data and makes it understandable at a glance.

- Try a food such as gelatin dessert or cake from a mix instead of a soft drink.
- Try four or five different colorings and find out what colors are the most popular.
- Try different intensities of the same color with the same drink and have the drinks rated for sweetness.
- Try different presentations of the same food such as cheese dip or salsa and rate them for taste. (Subjects would not know they are the same.) One sample might be presented in a classy dish, artfully garnished with parsley, while the other is served in an old plastic food container.
- The Color Matters experiment assumed that the food coloring itself had no effect on mouth-taste. Find out if this is a fair assumption by doing eyes-closed taste tests with soft drink or water samples colored as you had them in the original experiment.

Vocabulary and concepts appear on page 194.

"When choosing between two evils, I always like to try the one I've never tried before."
—Mae West

Color Matters Experiment

Purpose:
The purpose of this experiment is to find out _____
_____

Prediction:
_____
_____

Color Formulas:

A:_____   B: _____

Data:

Subject	Flavor Ratings		Appearance Ratings	
	A	B	A	B
1 (A first)				
2 (B first)				
3 (A first)				
4 (B first)				
5 (A first)				
6 (B first)				
7 (A first)				
8 (B first)				
9 (A first)				
10 (B first)				
11 (A first)				
12 (B first)				
13 (A first)				
14 (B first)				
15 (A first)				
16 (B first)				
Total:				
Average: (= total/number of subjects)				

Sources of Error:
_____
_____
_____
_____
_____
_____
_____

Conclusion:
_____
_____
_____
_____
_____
_____
_____
_____
_____
_____
_____

23. Memory Mystery

Have you heard the joke about the wall? You'd never get over it.

The human brain is a wondrous organ. Though it has been studied for many decades, we still have zillions of questions about how the brain forms and stores memories. You can join in on the research. Begin by asking a small, testable question. For example, "Does touching an object help you remember that you saw it?" Do the following experiment to answer this question, then go on to design your own memory experiments.

Psychology and Beliefs / Memory Mystery

Materials

- 2 boxes with lids such as shoe boxes (or two trays with dish towel covers)
- 20 different, small, common objects that will fit ten-in-a-box, such as tape, a comb, scissors, a cotton ball, a spoon, a rock, a candy bar, a pack of gum, a marshmallow, a balloon
- Stopwatch or a watch with a second hand
- Pencils
- Use of a copy machine or a computer and printer

I. Memory and Touch Experiment

"Character consists of what you do on the third and fourth tries."
—James Michener

Activity time: a few hours

The big picture is that you will have people, your "subjects," look at ten common objects for 40 seconds; then, with the objects out of view, they will write down the items they remember. Each subject will repeat this for a second box. With one box, the subjects will *look at and touch* the objects, while with the other box, the subjects will look at the objects but *not touch* them.

1. Figure out how you will find lots of subjects for your experiment. Between 8 and 16 subjects would be fine. The more subjects you have, the more reliable your results will be. You do not have to test all your subjects on one day or even in one week.
2. Write your prediction of what this test will find. Do you have a hypothesis, a possible explanation, behind your prediction? If so, record that too.
3. Create a form that each subject will complete. It will look something like this:

MEMORY FORM   **Subject #_____**

Part A Using Box ____

You are going to be shown ten objects. You are to **look at the objects but not touch them**. You will have 40 seconds to examine them. Then they will be covered and you will pick up a pencil to list the objects you saw. You will have 60 seconds to complete your list.

1 _____   6 _____

2 _____   7 _____

3 _____   8 _____

4 _____   9 _____

5 _____   10 _____

Part B Using Box ____

You are going to be shown ten objects. You are to **look at the objects** and **touch them** without picking them up. You will have 40 seconds to examine them. Then they will be covered and you will pick up a pencil to list the objects you saw. You will have 60 seconds to complete your list.

1. _____   6. _____
2. _____   7. _____
3. _____   8. _____
4. _____   9. _____
5. _____   10. _____

4. Prepare your boxes and label them "X" and "Y." Label the sides as well as the top because the tops could get mixed up. To test more than two people at a time you can make more than one set of boxes. No item should be so small that it gets hidden by the others. Create a testing station by putting your prepared boxes on a table. If you have duplicate boxes, arrange the items inside so that they are in the same positions.

5. You need to keep the testing conditions for your two boxes fair. Future subjects cannot be allowed to look over the shoulder of subjects taking the test. Since people tend to do better after practice, the second box gets a scoring advantage. To make sure that an "order effect" does not occur, have half your subjects do the touch procedure first and the other half do the "no touching" procedure first. You also want half your subjects to do box X as the touching box while the other half does box Y as the touching box. This is

because one group of items may be naturally more memorable than the other group, and we don't want this to skew the results either. If you arrange things as follows, any advantages apply equally to both boxes and both observing methods, so their effects cancel out.

1/4 Subjects Trials # 1, 5, 9, . . .	1/4 Subjects Trials # 2, 6, 10, . . .	1/4 Subjects Trials # 3, 7, 11, . . .	1/4 Subjects Trials # 4, 8, 12, . . .
Do Part A, *don't* touch, first with box X Code: AX	Do Part A, *don't* touch, first with box Y Code: AY	Do Part B, *do* touch, first with box X Code: BX	Do Part B, *do* touch, first with box Y Code: BY

6. On each subject's form, fill in the letters of the boxes and circle the section they are to do first according to the table. Have subjects stand in front of the proper closed box, form and pencil within reach. Read the directions aloud as they read along with you. Make sure that they understand what they are to do.
7. Have them open the boxes. Start the timer.
8. Close the boxes yourself after 40 seconds.
9. Allow 60 seconds for them to write what they remember. Then say, "Pencils down. Thank you for your help!" Collect the forms.
10. Repeat this process until you have tested all subjects.
11. Score each test by counting all the correct answers. For example, if a subject listed eight items but two of them did not actually appear in the box, her score would be six.
12. Find the average scores for the "don't touch" and "do touch" boxes by adding up each set of scores and dividing by the number of subjects.
13. Count the number of subjects who scored higher by each method. Turn your counts into percents by dividing them by the total number of subjects and multiplying by 100.
14. Think about how your experiment was carried out. Were there imperfections that could affect your results? Maybe some subjects got more time. Maybe there was an interruption during some tests. These things should be listed under "Sources of Error" on the record sheet.
15. Make two bar graphs to make your results visual. You will have one graph for the average scores, and one for the percent of subjects who scored higher by each method. See the Graphing Appendix on page 204 for help with making a good graph.
16. Compare the two averages and the percents. Any trend? Write your conclusion considering the sources of error that you noticed.

II. Design Your Own Experiment on Memory and Learning

Activity time: a few hours

You've seen how experiments are designed.

- Find a variable you want to learn about.
- Test it two different ways.
- Keep everything else the same.
- Do lots of trials.
- Collect data in number form, if possible.

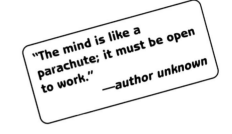

"The mind is like a parachute; it must be open to work." —author unknown

Apply these rules to design your own experiments. Here are some ideas but your own ideas may be the best ones of all.

For remembering objects in a box, these other questions could be asked:

- What's the effect of the age of the subjects? Compare results of different age groups.
- What's the effect of the gender of the subjects? Compare results of males vs. females.
- What's the effect of seeing objects vs. seeing the name of the object in writing only?
- What's the effect of seeing objects vs. seeing objects labeled with their names?
- What's the effect of a time lag between seeing them and listing them? Compare no lag to 10 seconds, 20 seconds, 60 seconds.*
- What's the effect of noise? Compare results collected in silence vs. those collected with relaxing music. Or silence vs. the sound of hammering.
- What's the effect of standing up vs. sitting down?
- What's the effect of knowing a reward will be earned for a perfect score?

For remembering digits in a long number, ask these questions:

- What's the effect of seeing the number vs. hearing the number?
- What's the effect of seeing a series of digits vs. a series of dice faces?
- What's the effect of seeing the digits evenly spaced vs. seeing them chunked together in two's or three's?

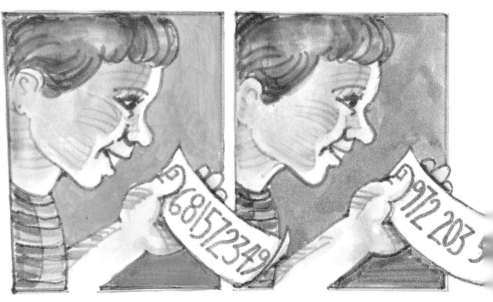

* The data from this study should be made into a *line* graph if more than one time lag is investigated.

- What's the effect of noise?
- What's the effect of the age of the subject?
- What's the effect of the gender of the subject?

For remembering a sequence of words, you could ask any of these questions:

- What's the effect of the words being in black vs. in color?
- What's the effect of age?
- What's the effect of gender?
- What's the effect of seeing the words vs. hearing the words?
- What's the effect of a time lag between seeing them and listing them?
- What's the effect of noise?

For remembering a sequence of pictures, these questions might bring out some interesting answers:

- What's the effect of seeing the pictures in black and white vs. seeing them in color?
- What's the effect of seeing pictures of objects vs. seeing pictures of objects labeled with their names?
- What's the effect of age?
- What's the effect of gender?
- What's the effect of noise?
- What's the effect of a time lag between seeing them and listing them?

For remembering an odor or a sequence of odors, try asking one of these questions:

- What's the effect of a time lag between smelling them and matching them?
- What's the effect of age?
- What's the effect of gender?
- What's the effect of knowing a reward will be earned for a perfect score?

The data from most of these studies are perfect for bar graphs. See the Graphing Appendix on page 204 for help with making a good graph.

Vocabulary and concepts appear on page 195.

Memory Mystery Experiment
Does Touching Objects You See Improve Your Memory of Seeing Them?

Purpose:
The purpose of this experiment is to find out _____
_____

Prediction:
I predict that _____
_____

Hypothesis:
I made that prediction because _____
_____

Data Table:

NUMBER OF ITEMS REMEMBERED

Procedure Code	Subject	Looking Only	Looking and Touching
AX	1		
AY	2		
BX	3		
BY	4		
AX	5		
AY	6		
BX	7		
BY	8		
AX	9		
AY	10		
BX	11		
BY	12		
AX	13		
AY	14		
BX	15		
BY	16		
Total:			
Average Score: (total/number of subjects)			
Number of Subjects with Higher Score:			
Percent of Subjects with Higher Score:			

Sources of Error:
_____
_____
_____
_____
_____
_____
_____
_____
_____

Conclusion:
_____
_____
_____
_____
_____
_____
_____
_____
_____
_____
_____
_____
_____
_____

My Own Memory Experiment

Purpose:

Prediction:

Hypothesis:

Data Table:

Sources of Error:

Conclusion:

24. Snack Foods
What Controls Your Taste?

What are your favorite cereal, cola, and cheese snack brands? Are you influenced by the name, packaging, or advertising? Soft drink and snack food producers seem to think you are. They feature fun-loving, adventurous, young people in their ads with the idea that if you like the image, you'll buy the product. But aren't consumers smarter than that? Don't we choose foods and other products based on the actual product? Do this experiment to find out if product image affects people's judgment of taste.

If Barbie is so popular, why do you have to buy her friends?

Materials

- A popular name-brand snack, such as cheese snacks, chocolate chip cookies, soft drinks, corn chips, or cereal
- A generic, grocery store, or little-known brand that looks about the same as the name-brand product
- Cups or napkins, depending on the product you choose
- Use of a copy machine or computer with printer, if possible

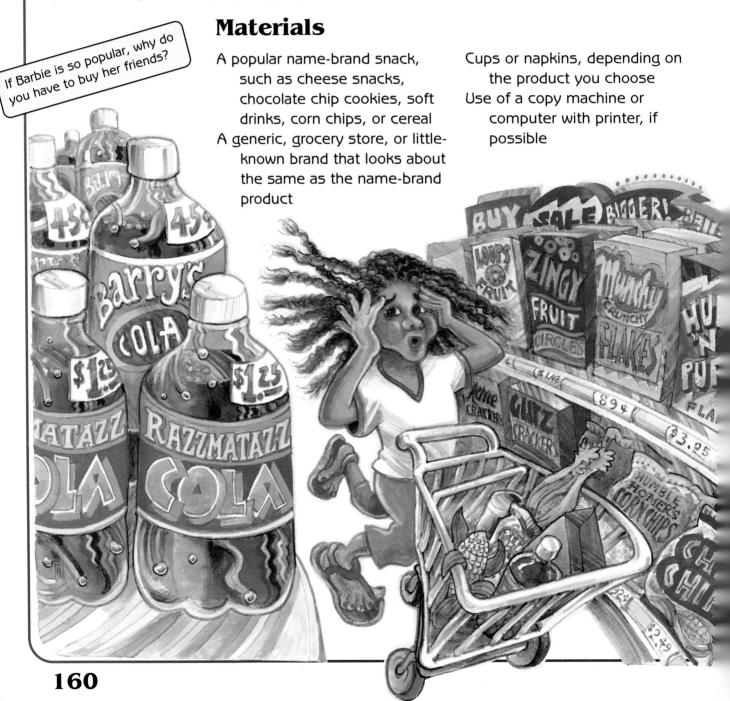

160

1. Testing for Image Influence

Activity time: a few hours

The big picture is that you are going to ask people, your "subjects," to taste two brands of a snack product and to rate how well they like them. The packaging, including the product name, will be showing. In reality, both products they eat will be the *same* brand. If people base their judgment only on the taste experience, then the ratings people give the products should be about the same. If the brand image does affect their judgment of the taste experience, the average ratings for the two products will not be so close. The bigger the difference, the more influence image had on subjects' judgment.

1. Choose two snack products that look similar, one a popular brand name and the other a generic, grocery store, or little-known brand that looks about the same as the name-brand product.
2. Figure out how you are going to get lots of subjects to take your taste test. People love to test snack foods so it shouldn't take much arm twisting. Is there a club meeting, party, or other event coming up soon where you could set up a little booth? Between 6 and 15 subjects would be fine. But the more subjects you have, the more reliable your results will be.
3. On the record sheet, write a clear purpose of the experiment and your prediction of what this test will find.
4. Create a form that looks something like this:

For each brand, taste a sample and rate how well you like it by circling one number.

Brand _____ (Fill in when you make the form)

   1   2   3   4   5   6   7   8   9   10
tastes very bad   tastes very good

Brand _____ (Fill in when you make the form)

   1   2   3   4   5   6   7   8   9   10
tastes very bad   tastes very good

Make as many copies of the form as you think you might need.

> **Note:** If you can't use a copy machine or computer and printer, you can make the forms by hand. An alternative is to verbally collect the ratings and record them on a data table on the spot, but this is not quite as good. You have to keep the data table hidden as you work so that one person's response has no influence on the next person. Also, there is no way to catch any mistakes you might make. With written forms, you can double- and triple-check your numbers.

5. Set up your testing station so that the original packaging is plainly visible. If you have chips or cookies, they could be in a bowl behind the package. Soft drinks could be poured into small paper cups in front of the bottles. Although you must have the packaging for both products, you will actually put out only one of the two products for tasting. Do all your preparation in private so you don't give away information that would hurt your results.

6. You must keep the testing conditions for your two samples fair. Be sure both samples are the same temperature and in the same kind of containers. Sometimes a food tasted first or last in a taste test has an advantage. To make sure that an "order effect" does not occur, have half your subjects taste one "brand" first and the other half taste the other one first. Then, if there is an advantage to being first or second, both brands will share it.

"Instead of thinking about where you are, think about where you want to be. It takes twenty years of hard work to be an overnight success."
—Diana Rankin

7. When you have finished collecting forms from all your subjects, clean up!
8. Copy the ratings onto the data table.
9. For each brand, add the ratings up and divide by the number of subjects to find the average. Compare the two average ratings.
10. Also count the number of subjects who rated each brand higher. Turn your count into a percent by dividing it by the number of subjects and multiplying by 100. For example, if 5 people out of 15 rated the famous brand higher than the other:

$$5 \div 15 = 0.333 \qquad 0.333 \times 100 = 33.3\%$$

11. To give your results visual impact, make bar graphs of the average ratings of the two brands and of the percent preferred values. For information on making bar graphs, turn to the Graphing Appendix on page 204.
12. Take a hard look at any weaknesses your study had. Did you eliminate possible order effects? Anything that kept your test from being fair should be listed as a source of error on the record sheet.
13. Did image affect judgment? Keep your sources of error in mind as you write your conclusion.

II. Design Your Own Study

Here are some ideas:

- See if the age of the subject is connected to how much image affects taste.
- See if the gender of the subject is connected to how much image affects taste.
- See if brand image affects people's judgment of clothing or other products.

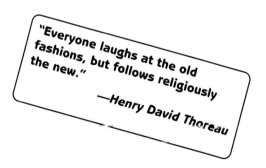

"Everyone laughs at the old fashions, but follows religiously the new."
—Henry David Thoreau

Vocabulary and concepts appear on page 197.

Snack Food Experiment

Purpose:
The purpose of this experiment is to find the effect of _____
_____

Prediction:
_____
_____

RATINGS OF THE SNACK FOODS

Subject	Famous Brand: _____	Other Brand: _____
	Really both samples were the _____ brand	
1		
2		
3		
4		
5		
6		
7		
8		
9		
10		
11		
12		
13		
14		
15		
16		
Total:		
Average:		
Number of Subjects with Higher Score:		
Percent of Subjects with Higher Score:		

Sources of Error:
_____
_____
_____
_____
_____
_____
_____

Conclusion:
_____
_____
_____
_____
_____
_____
_____
_____
_____
_____

VOCABULARY AND CONCEPTS

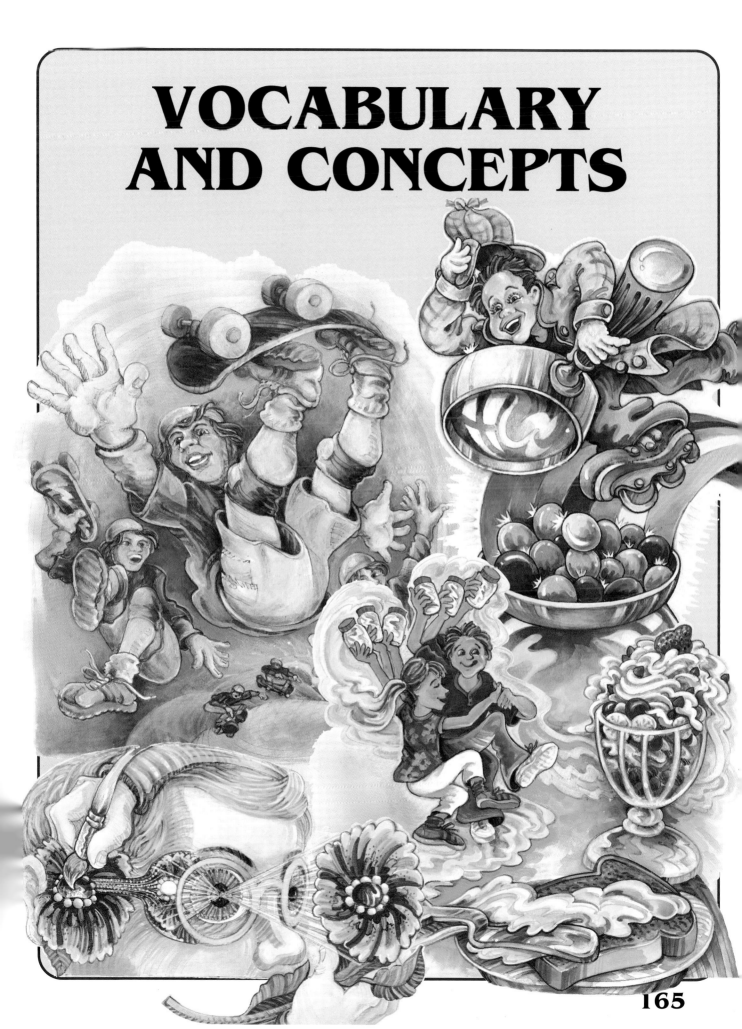

165

1. Mirror Mirror

Vocabulary

Receptors: special cells that react to sensory input by sending nerve messages to the brain.

Sensory impulses: nerve messages that go to the brain.

Proprioceptors: nerve cells in muscles, joints, and tendons that detect stretch. This information is sent to the brain, where it is translated into a sense of body position.

Kinesthetic sense: sense of body position and movement.

Motor impulses: nerve messages that go to the muscles to direct motion.

Eye-hand coordination: performing a motor task from visual and kinesthetic information.

What's Going On?

Your drawings look like chicken scratchings! What's your problem?

Just kidding. Everyone struggles with this because it's hard to do. When you look in the mirror, right and left are normal but forward and backward are opposite. What your eyes tell you to do is different than what your hand knows from experience. Curves and diagonal lines can be the hardest to get right as you work against habit when moving your hand closer to you and farther away.

You get two sets of cues during writing, one set from your eyes and one from your hand. You see where your pencil is going while tracing and can correct errors as you go. Continuous feedback from the eyes helps keep you on track. The other feedback is from your kinesthetic sense. Proprioceptors in your muscles and tendons allow you to feel how and where your hand moves. These senses work together very smoothly when you write normally. But when writing while looking in the mirror, these cues are in conflict with one another. The visual cues are *not* paired normally with the kinesthetic sense, and the result is not smooth. It takes great concentration to come up with the proper motion.

As you learn any new skill, new sensory and motor pathways are laid down in the brain. Repetition improves the speed and ease of mirror writing just as it did when you first learned to write as a child. Being a beginner at a skill takes time, effort, and concentration. Improvement requires repetition. Do you want to be good at piano, basketball, or math? The more you practice, the easier they become.

You write your name so often that you can do it well without visual cues. But when you watched this action in the mirror in the "Mirror Writing Introduction" your eyes told you that you were doing it all wrong. Conflicting cues made it hard to write as smoothly or quickly as usual. Part of you wanted to react to the visual cues, while part of you knew your hand could do it alone just fine.

An Historical Note: *The famous inventor and artist Leonardo da Vinci (he painted the Mona Lisa) wrote backwards in his big notebooks of ideas and designs. He wrote normally when communicating with others but his notes describing his ideas have to be held up to a mirror to be read. It's not known whether he was trying to protect his ideas from criticism, he was trying to make it hard to steal them, or, as a lefty, he just hated smearing the ink. Maybe he just enjoyed the novelty of writing backwards.*

2. Flavor Detective

Vocabulary

Blind test: when a subject does not know information that could affect his observations. In this study, the subject was "blind" to the color of the jellybean. He could not see and was not told the color of the jellybean. If the subject did know the color, the power of suggestion of the color would make it hard for him to give the jellybean's taste a fair test. A blind test doesn't have to be related to vision.

Trial: one try in an experiment. You did ten taste trials using your nose and ten trials not using your nose.

Experimental group: the group of trials you are particularly interested in. In this study the "no eyes, no nose" trials form the experimental group.

Control group: the group of trials needed for comparison. In this study, it is the "no eyes, using nose" trials. The control group allows you to evaluate the experimental group. Scientists don't assume to know what "normal" is. They collect data to be sure.

What's Going On?

Smell plays a very big role in the experience of taste. Taste and smell are the two chemical senses. When you taste or smell something, special cells have detected actual particles of the substance on your tongue or in the passages above your nose. Cells in your taste buds on the tongue can only respond to five tastes: saltiness, sweetness, sourness, bitterness, and meatiness, also called umami. Cells in the nasal passages, however, can respond to thousands of different molecules. When your brain gets messages from the mouth and the nose at the same time, it blends the two chemical sensations, along with texture and temperature sensations into one flavor experience.

Taste buds detect the sweetness of the jellybean, but the molecules that make one flavor lime and another orange are detected in the nose. In fact, what we experience as taste is *mostly* smell. When your nose is plugged up by holding it or by having a cold, food is pretty tasteless. Without smell, even onion and apple taste the same. (Try it!)

A common source of error in this activity is letting some air move into the nose during swallowing. When this happens, flavor detection is improved, and your score is unfairly raised. The opposite error occurs if a subject has a cold. Then the percent correct when not holding the nose will be lower than normal.

This activity is a good one for learning how to design experiments. An experiment is well designed if it tests something *fairly*. Fairness is a simple idea, but it makes for strict rules about what you can and can't do when experimenting. Remember how you picked colors out of the cup without looking at them? That improved fairness by reducing the chance that the subject could guess what colors you would give him. He may know you well enough to guess correctly some of the time. You don't want that to affect your results, so you don't choose the colors, you let chance choose them.

Another way the experiment's design insured fairness was where the jellybean-giver could not give any feedback to the subject until the end of the test. If you say "Yes, that was orange," the subject knows the next one is less likely to be orange. No fair! In high-stakes experiments, such as those for testing new medicines, the people giving the test are "blind" to information that could affect the results, just like the subjects. For example, the person handing out the medicine and recording the results only knows that one group of subjects get medicine A and another group gets medicine B. Then there is no way she can accidentally give clues to the subjects about what should be happening. A third person, the mastermind of the study, knows which medicine is the real medicine and which is fake. This kind of setup is called a *double-blind test*.

These aspects of experiment design show some small ways tests are kept fair. More obvious ways include doing lots of *trials* and having a *control group*.

In this experiment, two of you each did ten trials, with and without holding your noses. Data collected from ten subjects doing that would be more convincing than data from two subjects. Data from 100 subjects would be more convincing yet. The more trials, the better, until you run out of time or money.

Your control group was the data collected with the subject's nose open (and eyes closed). Through this part of the study, you found what is normal for flavor identification using nose and mouth. You might have assumed identifications this way would be 100% correct, but likely, they were a little less than 100%. What percent it was affects how you look at the percentage for the nose-closed results. If you don't have control group data, you can't fairly judge the rest of the data. Let's say you find that the nose-closed score is 83%. Your conclusion about the effect of smell is still unknown. It depends on what the score *with* the nose was. If, with the nose, the score was 84%, you would conclude that nose doesn't seem to matter. If it was 97%, then the nose made a pretty big difference. Having a control group, a solid basis for comparison, is essential.

3. What You See Is What You Get . . . Or Is It?

Vocabulary

Reception: what your senses get from the world; sensory input.

Perception: what you *think* you get from your senses; the brain's interpretation of sensory input.

Retina: the light-sensitive nerve tissue lining the inside of the eyeball.

Optic nerve: the bundle of nerve cells that connects the retina of each eye to the brain.

Blind spot: the location in the back of the eye where the nerves of the retina come together to form the optic nerve to the brain. The two blind spots (one in each eye) result in areas of blindness (usually unnoticed) in the field of vision. These areas of blindness may also be referred to as "blind spots."

What's Going On?

Eyes are totally useless without light. The part of the eye that reacts to light is the lining inside the eyeball at the back. It is called the retina. The retina contains nerves and blood vessels as well as special cells that respond to light. The light-sensitive cells in the retina send electrical messages through nerves to the brain. The brain translates the messages into vision.

The nerve pathway from the eye to the brain is called the optic nerve. The nerve cells in the retina come together in a bundle to form this nerve. At the point where the optic nerve exits the back of the eye, there are no light-receptive cells to send vision messages. This is what causes the blind spot. The optic nerve exits the back of the eye level with the center to the side toward the nose. This produces a gap in your field of vision that is on the same side as your ear. When you turn the card so the spot is below the triangle or in any position other than to the outside, the spot on the card never disappears. In those positions, it cannot coincide with the blind area.

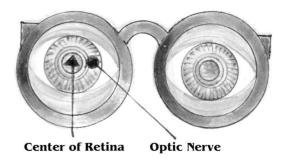
Center of Retina **Optic Nerve**

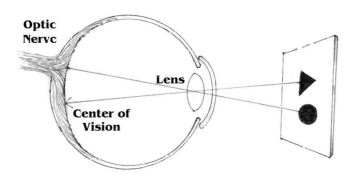

The eye–brain teamwork is so effective that you never notice that there is an area in your field of vision that is missing. Your brain fills in the gap. You perceive more of whatever surrounds the blind area. If you did the blind spot test with colored paper, you noticed that while the spot disappeared, the color of the area matched the color of the paper you used. With a line drawn along the card, the spot disappears but not the line. The brain doesn't seem to like blank or nonsense input. It turns the missing area of vision into something that fits the rest of the picture.

While we perceive vision where there is none at the blind spot, it is just the opposite with the blood vessels. The blood vessels are always there in front of the light-sensitive cells. Light that reaches the back of the eye has to pass through the layer of tissue that carries the blood vessels before getting to the light-sensitive cells. We don't usually see the vessels though because the light-sensitive cells only send impulses for images that *change*. Since the blood vessel positions do not change, the brain never gets messages about them. This is a good thing because it would be very annoying to see blood vessels everywhere you looked! In this activity, the flashlight casts moving shadows of the blood vessels that stimulate the retina. The retina sends impulses about the shadows and you see the shadows as darker areas on the paper.

4. Depth Perception

Vocabulary

Depth perception: the ability to judge how near or far away objects are.
Binocular vision: seeing with two eyes. This allows for good depth perception.
Monocular vision: seeing with one eye. Depth perception is not very good. (Telescopes are monocular, while binoculars are, well, binocular!)
Vision therapy: training in using the eyes to improve coordinated vision.

What's Going On?

You probably noticed in looking at the can or bottle that each of your eyes has a slightly different view of the objects you see. When you look at an upright soup can, your right eye sees a little more of the right side of the can, while your left eye sees more of the left side. The closer the can is to your eyes, the bigger the difference between the two views. As an object moves farther away, the views of the two eyes become more and more similar though never exactly the same.

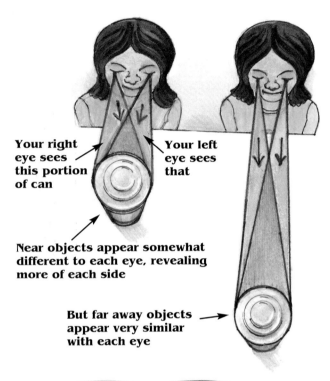

Your right eye sees this portion of can

Your left eye sees that

Near objects appear somewhat different to each eye, revealing more of each side

But far away objects appear very similar with each eye

View with left eye

View with right eye

Your brain recognizes this relationship between the views of the two eyes and the distance from the object. It processes the information almost instantly. You don't have to think about the distances to objects; you know where they are. You easily reach to the right position to shake someone's hand or to catch a ball. This is the advantage of binocular vision. While our survival today doesn't depend on catching or batting, depth perception for our ancestors was critical. Hunters need to be able to pinpoint their prey. Prey animals, as well, need to be able to run and jump, correctly judging the shape of the land to avoid obstacles.

If someone only has one working eye, judging distance is possible to a degree. Distance cues come from sizes and angles, shadows, and what's covering what. Moving the head from side to side adds some of the information that binocular vision provides. While helpful, these cues are not enough for reliable depth perception when objects move quickly. Activities with moving objects or that require quick reaches for objects in different positions cause trouble for a person with monocular vision. This includes driving, waiting tables, and playing baseball. A person with monocular vision can be a wonderful athlete in sports such as swimming, running, and weight lifting but won't be too great at fencing or most sports with balls.

It is not uncommon for a person to have two healthy eyes yet depend mostly on just one. An eye doctor can detect this problem and provide vision therapy. Vision therapy trains the person to use both eyes more equally, and improves depth perception as a result.

5. Where Did I Put My Toes?

Vocabulary

Receptors: nerve cells that react to sensory input by sending messages to the brain.
Proprioception: the collection of information from internal receptors that results in awareness of body position.

Proprioceptors: receptor cells in muscles, joints, and tendons that detect stretch. This information is sent to the brain where it is translated into a sense of body position.

Center of gravity: the point around which the weight of an object is evenly distributed.

Semicircular canals: organs of balance in each inner ear that detect head rotation.

Kinesthetic: related to the sense of body position and movement.

What's Going On?

We are usually unaware of all the nerve activity keeping us in control of our body position and motion. But the activity is quite complex. Vision, touch, the inner ear, proprioceptors throughout the body, and memory all play important roles. The brain processes the input and directs the muscles appropriately. For healthy people, no conscious effort is needed to coordinate all the processes. Injury and disease, however, can damage any part of this system and lead to problems controlling motion.

In the case of Ian Waterman, a virus destroyed the touch receptors in his skin and the proprioceptors in his muscles, tendons, and joints. Below his face, he has no feeling, and he can't tell where his body parts are except by looking. The nerves directing muscle motion were unharmed. The fact that the damage was complete yet at the same time limited to touch and proprioception makes Mr. Waterman's case very interesting to researchers trying to understand more about what is sometimes termed the "sixth sense." Over the course of three difficult years, he relearned to control his movements using only his eyes for feedback. He learned to sit up, feed himself, walk, and do all the other motions we take for granted. Yet it all takes a great deal of effort.* He never gets to relax. For example, a cup of coffee in a Styrofoam cup is a hazard for him. He can't feel the strength of his own grip. If he holds it too loosely, he'll drop it. If he grips it too tightly, he'll crush it. The solution? Wait until the coffee has cooled and then watch very closely. That's a lot of effort!

His case helps us understand normal function. Let's say you are sitting in a chair. You feel your feet touching the floor. You feel your rear end pressing on the chair seat, and you can tell if the seat is soft or hard. So just from touch, you have evidence that the soles of your feet and your seat are pointing down. But you process more information without consciously noticing it.

Your limbs and every body part send information on their location to the brain. Muscles, joints, and tendons have receptor cells, called proprioceptors, that detect the structure's degree of stretch. The brain uses this information to determine the position of the body part.

The bone-encased inner ear has organs of balance that tell your brain about head position, head motion, and which way is up. In addition to the cochlea, an organ of hearing, the inner ear is composed of semicircular canals and a connecting area that contains two small sacs, all organs of balance.

The two sacs contain tiny crystal rocks stuck to the tips of thousands of flexible hair

---

*If you did mirror writing in "Mirror Mirror" on page 10, you might be interested to know that Mr. Waterman would have no trouble tracing images in a mirror. You have trouble because your proprioceptive cues conflict with your visual cues. He only gets the visual cues, so there is no conflict.

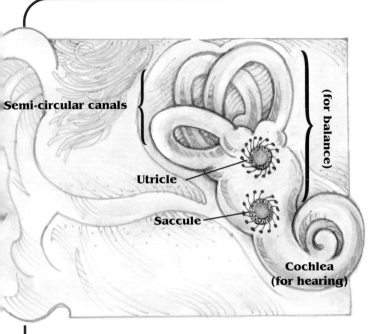

cells. When the head tilts or accelerates, the rocks bend the hairs, triggering impulses to the brain. The semicircular canals are three fluid-containing loops oriented in perpendicular planes. When the head moves, the fluid moves hair cells that trigger signals to the brain. Since the fluid movement is different in each plane, three-dimensional movement is perceived as a result.

These systems work perfectly for most motions. But they don't work well when you spin. You feel dizzy after spinning because, when you stop, the motion of the fluid incorrectly signals spinning in the opposite direction. The eyes dart in the direction of the perceived turn, and pointing drifts in that direction as well. Visual cues don't match the information from the ear and throw off your ability to control your motion. You stumble or fall over. You can't tell which way is up. The mixed messages you get from the eyes and the inner ear also trigger the woozy feeling of motion sickness.

For normal motion control, your brain adds up all the information it gets and draws a mental picture of where all your body parts are. The overlap of information refines the picture. Your memory directs a needed motion from this starting point. With eyes closed, unpracticed motions are a bit clumsy when done off to the side but pretty accurate when done in the middle. Yet you are not always perfect in matching a fingertip to an elbow. With eyes open, you get more feedback to make more exacting moves. But in total darkness, most people can still walk, sit, feed themselves, and write their names easily. We know the kinesthetic patterns by heart and don't have to see to do them. Mr. Waterman, however, would be helpless in total darkness. For him, a power outage can be a great fright.

The activities in Part II may have made you feel uncoordinated as well. They are, with the possible exception of number 6, impossible. *Your center of gravity must stay over your base of support or you will fall.* The activities seem like normal motions but are set up so that you are not able to adjust your weight to balance yourself. Countless times a day you shift your weight distribution without thinking, having learned the moves when you first learned to walk. In Part III, you observed the body's automatic adjustments to weight shifts. Either you moved your center of gravity over your base or you moved your base under your center of gravity. You must do one or the other or you fall!

When you bend forward, you must move your rear end backward to keep your center of gravity over your feet. When you go up on your toes, you must first shift your weight forward to put your center of gravity over the toes. When you stand from a seated position, your base shifts from your seat to your feet, and you must either get your weight over your feet *or* slip your feet under where your weight is. Lifting your right foot is similar to going up on tiptoes. First you must shift your weight to a position above the new base. You move to the left, so your center of gravity is over your left foot.

Whether or not you could knock over the video depends on your body weight distribution. Having lots of upper body weight makes it impossible. Someone slim

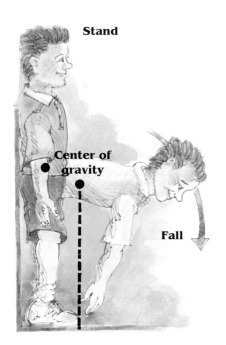

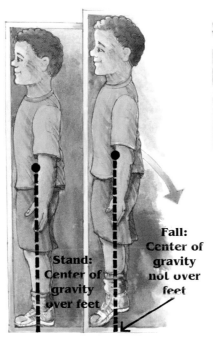

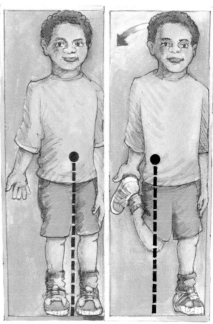

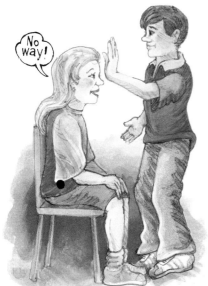

in the upper body and heavier in the lower body might succeed. For most people, though, if you lean forward enough to reach the video, your center of gravity moves forward to a point that is not above your knees, and you fall over. You probably never really fell because you could tell it was going to happen before it did, and you protected yourself. We are lucky that the brain can process all this stuff on auto pilot! One time that auto pilot may fail is when your limbs "fall asleep." If this has happened to you, you have experienced a partial loss of proprioception. If your legs or feet "fall asleep," you can't walk properly until the nerves return to normal.

6. Food Preservation I
Dried Apples and Fruit Roll-Ups

Vocabulary

Dehydration: loss of water.
Vitamin C: ascorbic acid. Vitamin C prevents the discoloration of fruit.
Oxidation: a reaction with oxygen. Fire, rust, and the browning of apples are all examples of oxidation.
Evaporation: a change of state from liquid to gas, as in drying.
Convection oven: an oven that has fans for continuous circulation of heated air.
Microbes: also called microorganisms. Microbes are microscopic (very small) organisms such as mold, yeast, and bacteria.
Mold: a group of fungi that often has a fuzzy appearance.

What's Going On?

Apples and other fruits are mostly water. The rest is sugars, acids, fiber, vitamins, and minerals. Decay is caused by the growth of microorganisms, which thrive in moist surroundings. If the water content of fruit is lowered to about 20%, the target level for the dried slices and roll-ups, the microbes do not grow, and the food lasts much longer.

Preparing the fruit in this activity speeds up dehydration in two ways. Water leaves the fruit by evaporating at its surface. Slicing the fruit or spreading it thinly increases the *surface area* of the fruit where evaporation can occur. The *heat* of the oven speeds the rate of evaporation because water particles absorb energy from the heated air and change to the gas state more readily.

Preserving food by any method involves preventing chemical reactions, especially the ones carried out by microbes. When food spoils, microbes are living and growing on the food, digesting it for their own use, and leaving behind possibly nasty wastes. Warmth, as well as moisture, promote their growth. Since we want our food for ourselves, we outsmart the microbes by making food too cold (freezing and refrigeration), too dry (dehydration), too salty (salt curing), too sweet (jams and jellies), or too acidic (pickles) for them. When we do this, the microbes can only grow slowly or not at all. We buy time to eat the food before they do. While microbes make the most drastic changes to foods, light and air can also reduce the quality of foods through chemical changes. Vacuum packing and the use of oxygen absorbers keep out air, while opaque containers keep out light to reduce the rate of such changes.

Science requires good records. The best gardeners, business people, gamblers, and chefs make sure they learn from their experiences. They take notes so they know what they did before and what the outcomes were. They analyze what has happened and plan their next effort accordingly. That way their chances for improved success increase with each new try. That's the reason for noting temperatures and times and labeling the bags of dried fruit. Let's say, a week after drying some apple slices, you find mold growing on them. If you kept good records, you can change your method in the future. You might choose to adjust the drying time, the temperature, or any other variable. If you didn't keep track, you won't remember what you did, so your experience was wasted. You are forever a novice, guessing and hoping for the best. With written records, progress is made.

7. Food Preservation II
Refrigerator Pickles

Vocabulary

Spores: dormant cells from fungi and bacteria that can develop into active fungi and bacteria under the right conditions.
Pickling: using salt and/or acids such as vinegar to preserve and flavor food.
Acid: having a pH less than 7. Acidic foods taste sour and include lemons and vinegar.
Sterile: microbe-free.
Sterilize: to make something microbe-free. In home canning, equipment is sterilized using heat. In other situations, microbe-killing chemicals or radiation may be used.
Diffusion: movement from an area of higher concentration of a substance to an area of lower concentration of the substance.
Osmosis: diffusion of a fluid across a membrane.
Cell membrane: the thin sac that surrounds all or most of a cell. In plants, a cell wall is found just outside the cell membrane.
Preservative: a substance that extends the useful life of a fresh product. Salt and vinegar are natural preservatives.
Dehydration: water loss.

What's Going On?

The whole idea of pickling is to make foods last a long time. The salt and vinegar in pickles keep microbes from growing and causing decay. Refrigeration slows microbe growth and also slows the chemical reactions that occur between oxygen and foods. None of your cucumber samples was sterile. All were exposed to the air, which carries a wide variety of fungal and bacterial spores. Given the right conditions, the spores grow and multiply. Your untreated room-temperature sample probably became colorful with mold and bacteria in just a few days. Think about this when you leave food on the table for a long time before covering it.

Years ago, before produce was routinely shipped all around the globe, people could only eat fresh cucumbers a few months of the summer. Cucumbers didn't keep very long in the cellar and when the last fresh ones were eaten or went bad in the fall, that was it until the next growing season. But cucumbers are easy to grow in large quantities and to keep them edible for a longer season homemakers "put them up." They pickled them. That way their cucumber crop could supply tasty food all year long. To make pickles last all year without refrigeration, the empty jars were sterilized and then the filled jars of pickle mixture were again boiled before sealing.

With fast shipping and refrigeration now available almost everywhere, vegetable growers can usually get their produce to a market while fresh. We no longer have as strong a *need* to make pickles to preserve the food. We still have them though because we like their crunch and zing. Food traditions are long lasting and the place of the mighty pickle as a food accent is secure.

The two pickle recipes in this activity are called refrigerator pickles because they have not been processed in boiling water. (It was safer for you to make them that way.) Instead of *killing* all possible microbes, the growth of any existing microbes is discouraged by the cool temperature of the refrigerator along with the preservative actions of the vinegar and the salt.

Something happened when you put salt on the cucumbers. Water oozed out of the cells where the salt was. This is because water naturally moves from where it is highly concentrated to where it is less concentrated. Since the water moved through cell membranes in the process, it is called *osmosis*. The salt you sprinkled dissolved in a bit of liquid on the surface of the cucumber slices. As the surfaces were very concentrated in salt but low in water, the water inside the cells of the cucumber moved outward. The salt dehydrated the cucumber cells. Salt does this to microbes too and is why most microbes cannot grow in it. The vinegar doesn't cause dehydration but the acid in it is harsh on microbes in other ways.

Maybe you've heard that you can die of thirst drinking salt water. It's true. The food you eat and drink travels through the stomach and into the intestines. Water moves through the wall of the intestines to be absorbed into the bloodstream where it can be delivered to all cells in the body. However, if a person drinks salt water, this process runs backward and water moves from the bloodstream into the intestines. The thirst and dehydration of body cells gets much worse. Ughh! Please pass the *fresh* water!

8. Enzyme Excitement in Garlic

Vocabulary

Chemical reaction: a rearrangement of atoms to make a new substance.

Enzyme: a complex protein that promotes a chemical reaction. The enzyme is not changed by the reaction.

Substrate: a substance acted on by an enzyme to produce a new substance.

Protein: a large molecule in a particular shape that is made of a long chain of smaller units. Proteins form enzymes and structures in cells of living things.

Denatured protein: a protein molecule that has had a change in its three-dimensional shape. If the protein is an enzyme, denaturing destroys its action.

Active site: a cozy spot in the enzyme molecule that the substrate molecule fits into perfectly. Once the chemical reaction is finished, the substrate molecule leaves the active site, and the process can repeat.

Vacuole: a sac-like structure within cells.

What's Going On?

The different treatments of the garlic cloves in this experiment result in different flavor and aroma strengths. Whole garlic cloves have very little aroma because the strong-smelling and strong-tasting substances are only produced when garlic is cut or

Food Science / Enzyme Excitement in Garlic

smashed. An enzyme is responsible for this chemical change. The enzyme, called *allinase*, reacts with an odorless substance called *alliin* and changes it into a new, strong-smelling molecule. Both substances were present in the microscopic cells of the garlic all along, but the enzyme is packaged in tiny sacs called *vacuoles*. The vacuoles keep it separate from the alliin.* When the cells are sliced or crushed, the vacuoles break, and the substances immediately combine. This process is similar to illuminating those light sticks where you have to break the little glass capsule to release a chemical that "turns on" the light. The alliin fits into a certain spot, the active site, on the allinase molecule. Its structure is changed, and a new substance called *allicin* is formed. Allicin has a strong smell and taste. Actually, allicin quickly changes into other flavorful and aromatic substances. Once it has formed, the floodgates have opened regarding garlic flavor.

All three of your samples were cut, so why weren't they all equal in producing aroma and flavor? Well, the allinase enzyme that starts the process is destroyed by heat. So the clove that was microwaved before crushing lost its supply of the enzyme before the enzyme ever got to take action. The garlic never became very flavorful. In the other two samples, the enzyme did take action when the garlic was crushed, so the strong allicin and allicin products did form, making the garlic much more flavorful. Of these two samples, a little flavor may have been lost as a result of cooking the third clove, but not very much. So I guess you have solved the mystery. The 40 cloves of garlic in the mild chicken recipe were not chopped before cooking. It should also make sense that coarse chopping of garlic will not produce as much flavor as crushing. Any cells not broken will never produce much flavor once they are cooked.

Realize there is no way to eat and taste raw, *uncut* garlic because as soon as you bite it the chemical reaction occurs, and allicin is formed. The best you can do is to put a whole clove in your mouth and move it around.

RECIPE FOR 40-CLOVES-OF-GARLIC CHICKEN

About 8 pieces of chicken
40 cloves of peeled, whole garlic
3 tablespoons of olive oil
1 cube of chicken bouillon dissolved in 1/2 cup hot water
Juice of one lemon
1 teaspoon dried thyme
1 teaspoon dried rosemary (smell the spices to see if you like them; they are optional)
Salt and pepper

1. Put the olive oil in a frying pan.
2. Peel the garlic cloves, toss them in the oil in the pan (no heat), and put them into a baking dish.
3. Brown the chicken in the remaining oil. Sprinkle it with pepper and a little salt.
4. Put the browned chicken into the baking dish on top of the garlic cloves.
5. Pour the bouillon and lemon juice over the chicken.
6. Sprinkle with the thyme and rosemary or spices of your choice.
7. Cover with foil and bake for 70 minutes at 350°F (177°C).

---

*Research in the 1990's: The Chemistry of Garlic, online at http://chemed.chem.purdue.edu/genchm/topicreview/bp/2organic/garlic.html.

The interesting habits of garlic don't end once the garlic has been swallowed. Perhaps you've noticed that even strong mouthwash can't always get rid of garlic odor. When a person eats a lot of garlic, some of the smelly substances survive digestion and travel through the bloodstream. They escape the body in sweat and in air exhaled from the lungs. The toothbrush and mouthwash just can't do the job!

While you may not like to smell like garlic, there is an up side to garlic that goes beyond the taste. Garlic has a growing reputation for its health value. Studies have shown that the strong-smelling compounds in garlic fight bacteria, fungus, and cancer. They also lower the risk of heart disease. Because these substances slow the clotting of blood, garlic chemistry has even led to breakthroughs in the creation of blood-friendly plastics used for human implants such as heart valves.[‡]

Did you notice that, when you came in from outdoors, the garlic smelled stronger than before you left? The fact that you quit noticing a smell after a while is called *odor fatigue* or *odor adaptation*. Your sensory cells for smell send few messages to your brain about "old news." Your brain also filters out what becomes background information. Once you move to a place free of the odor, you can smell it again when you return. The moral of the odor fatigue story is that if you haven't taken a shower for a few days *you* may not be able to smell yourself, but, very likely, other people can!

[‡] Interview with Eric Block, Ph.D., by Richard A. Passwater, Ph.D., published in Health World Online, online at http:/www.healthy.net/asp/templates/interview.asp?PageType=Interview&ID=173.

9. Testing Foods for Fat and Starch

Vocabulary

Fats and oils: a category of molecules found in living things, often used as a way to store energy. Fats and oils have similar chemical structures. A fat is solid at room temperature, while an oil is liquid.

Translucent: partly transparent but not clear.

Opaque: not at all see-through.

Evaporation: changing state from liquid to gas at the liquid's surface.

Starch: a category of molecules made mostly by plants. Starches are easy to digest for a quick energy source.

Carbohydrate: starches and sugars.

Positive test: the substance or condition being tested for is found.

Negative test: the substance or condition being tested for is not found.

False negative test: when the substance or condition being checked for is not found but is actually present.

False positive test: when the substance or condition being checked for is shown to be present but is actually not present.

What's Going On?

Fats, oils, and water are all absorbed by paper fibers making the paper translucent. You can distinguish fat from water because the water will evaporate in time, leaving the paper opaque, while the fats and oils will not. It's possible to get a false negative test if the fat content is low or is bound up with other food components. Fats will spread more if warmed, which is why the samples should be at least room temperature. It's possible to get a false positive test if a bit of another food touched or became mixed into the sample.

Fats and oils, also called lipids, supply more energy per gram (9 Calories) than any other nutrient. This is why they are efficient for food storage. Fats help insulate us from the cold and give us protective cushioning. Fats and oils also dissolve and carry the vitamins A, D, E, and K through the body. In moderation, fats are essential for good health. Fats are slippery so some foods such as butter, bacon, and salad dressing are easy to identify as high in fat. Not all fat sources are obvious, though. Olives, cheeses, whole milk, meat, fish, seeds, nuts, and chocolate are also high in fat.

The iodine makes a dramatic color change in the presence of starch. The iodine joins with the starch in a union called a complex. The absorption of light is changed enough by the union that the color is different. Starch testing is done on foil, not paper, because paper itself contains starch. You could get a false positive for this test if one food contaminated another. A starch test on a food like beets is hard to judge because it already has a dark color. Probably the best way to test dark foods is to put several drops of iodine onto the foil then add a small amount of the food.

Both starches and sugars are carbohydrates, but starches are called "complex carbohydrates" because they are made of chains of sugar molecules linked together. They provide 4 Calories of energy per gram compared to fat's 9 Calories. That's why you can eat more starch than fat without gaining weight. Potatoes, rice, breads, crackers, and cereals are high in starch. For a ready supply of quick energy, athletes usually eat high-starch diets.

In the detection of many medical conditions such as strep throat, pregnancy, and the presence of HIV, tests like these are used to reveal a germ, a hormone, or an antibody. It's important to realize that a certain percentage of such tests (chemical or physical changes in a test medium or solution) are in error for some reason. A positive or negative test is an important clue, but the presence of the condition is not a certainty until additional tests and observations support the first.

10. Snap! Crackle! Popcorn!

Vocabulary

Embryo: an early stage of development; plant embryos are found in seeds. Much of the corn seed is filled with food to nourish the embryo during its first days of growth.

Seed coat: the tough covering of a seed.

States of matter: the physical forms of matter: gas, liquid, solid, and plasma.

Change of state: a transformation of a substance from one state of matter to another.

Optimum: the best possible.

Experimental group: the group of trials in which you are particularly interested. You have two experimental groups in this study: the soaked and the dried batches of popcorn.

Control group: the group of trials needed for comparison. The untreated batch of popcorn is your control group.

Germination: the sprouting of a seed.

What's Going On?

There is an optimum water content for kernels of popping corn. It is between 13.5% and 14.0%,* or about one-seventh the weight of the corn. This may be higher than you would guess from looking at those

---

*The Popcorn Board, Chicago, Illinois.

hard, hard kernels. If the water content is too high, the kernels pop more violently but end up small. Kernels that are too dry have a low popping rate, pop quietly, and also end up small. Popcorn producers adjust the water content of the kernels to produce the largest, tenderest puffs. If their product is too dry or too moist, it won't pop well, and customers won't buy it. To keep popcorn kernels fresh at home, store them in an airtight container in a cabinet rather than in the refrigerator. The refrigerator tends to dry them out.

Corn belongs to a very diverse species called *Zea mays*. *Zea mays* comes in six different main types: sweet corn, dent corn, flint corn, pod corn, flour corn, and popping corn. Each type has many varieties. Only popping corn varieties will pop. Native Americans in North and South America ate popcorn regularly and used it for hair and necklace decorations long before Europeans came to the New World. It was popped in a variety of clay containers or right on the cob. The planting of popcorn seems to have started in Mexico over five thousand years ago. Wampanoag people introduced English colonists to popcorn at the first Thanksgiving feast in Plymouth, Massachusetts.

The corn kernels you buy for popping have been processed only by drying. Adding water will trigger germination. You should see sprouts appear in about six days.

11. Boogie Woogie Butter

Vocabulary

Physical change: a change in matter that does not result in a new substance. Examples include melting, freezing, dissolving, and chopping.

Chemical change: a change in matter where the links between atoms are rearranged to make at least one new substance. The formation of rust and the burning of toast are examples.

Colloid: a mixture of one substance, often a liquid, with another substance in the form of tiny, nondissolved particles that never settle out. Colloids are halfway between a real *solution*, where one part of the mixture is dissolved in the other, and a *suspension*, where small particles are suspended for a while but settle out in time. Colloids may appear clear or cloudy but scatter light when a beam is shined through them. Milk is a colloid.

What's Going On?

Cow's milk contains water, vitamins, minerals, and all three of the substances that supply energy in foods: proteins, fats, and carbohydrates. It is the fat that is collected when butter is made, along with small amounts of the other components. Butter is about 90% milk fat. This fat is delicious. Some people like the taste of fat so much, they eat too much high-fat food, and this leads to weight gain and heart problems. These conditions have given fat a bad name. Yet some fat in the diet is necessary for life. A healthy diet contains some fat every day.

Cream is the fat-rich part of milk. It is less dense than the rest of the cow's milk so it

Create Your Own / Boogie Woogie Butter

floats to the surface when the milk is allowed to stand. It is skimmed or poured off the top and separated further into several products. Heavy whipping cream is about 40% fat. Light whipping cream is about 30% fat. Light cream is about 20% fat, and half and half is about 12% fat. The higher the fat content of the cream, the more butter you get and the shorter the time to make it. In contrast to cream, whole milk is about 3% fat and skim milk has almost no fat.

Shaking doesn't produce any new substances in cream, it just moves them around. So the change you observe is a physical change, not a chemical one. When you buy the cream, the fat is in tiny, invisible bits that are suspended by the water and protein. As you shake, the bits of fat stick to one another, growing in size until finally, they separate from the water and protein and become one big lump.

Before the 1930s, making butter at home was common. In the United States, before 1871 when the first commercial creamery started making butter in Iowa, making your own was the *only* way to get butter. Instead of shaking cream in a jar, a tool called a churn was used. The churn was a container that had some kind of paddle (a dasher) that hit the cream to force the fat globules together. Early European and American churns were made of wood and contained a dasher that moved up and down. Asian nomads churned butter in animal skin pouches. In the 1900s, most churns had glass jar bodies, metal lids, and a metal cranking device to move the paddles inside.

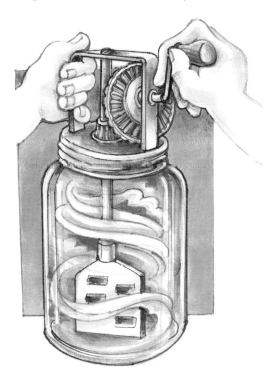

When you whip cream with a mixer, you add tiny air bubbles that force the tiny bits of fat to stick together around each bubble. This makes the cream firm. It helps to have your bowl and beaters cold because in warm temperatures the fat melts, losing its firmness. The higher the fat content, the more quickly your cream will become stiff. If you beat the cream too long, the fat all comes together as butter, which is disappointing if fluffy cream is what you had in mind.

12. Yogurt, Please

Vocabulary

Bacteria: simple, one-celled microorganisms. Bacteria are present in almost every environment.

Lactose: the sugar found in milk. The names of sugars end in "ose" such as sucrose (table sugar), fructose (fruit sugar), and maltose (malt sugar).

Lactose intolerance: the inability to digest milk sugar. People who are lactose intolerant get gas and pain in the intestines after eating certain milk products.

Protein: large molecules that form many cell structures and perform many functions. The main milk protein is called *casein*. It gives body to yogurt and cheese.

Carbohydrates: starches and sugars.

Denatured protein: a protein molecule that has been changed in shape.

Insulate: to slow the movement of heat from one place to another. Insulation keeps hot things hot and cold things cold.

Energy conservation: getting the most work out of a certain amount of energy.

What's Going On?

Bacteria are the tiniest of one-celled organisms. You may be used to thinking of them as enemies because some of them cause tooth decay and infections. But bacteria do good work in our lives as well. Certain bacteria capture nitrogen from the air and change it into a form plants can use for growth. Other bacteria replenish the soil by releasing nutrients from dead plants and animals. (See "Kitchen Compost" on page 103.) Some bacteria destroy pollutants. Others make important medicines. In food production, bacteria are responsible for making cheese, sour cream, some pickles, and sauerkraut. And, of course, they are the super-heroes of yogurt making.

Milk contains proteins, carbohydrates, vitamins, minerals, water, and, except in skim milk, fat. When yogurt is made, the sugar and the protein undergo changes. The *Streptococcus thermophilus* and *Lactobacillus bulgaricus* bacteria use the milk sugar, lactose, for energy to divide and grow.

They turn the lactose into lactic acid. Acids taste sour, so this changes the flavor of the milk. The acid also changes the shape of the protein molecules (denatures the protein), which is what makes the milk thicken to a semisolid. 110°F (43°C) is ideal for the action of the bacteria. When the milk is refrigerated, the growth of the bacteria slows.

Refrigeration is interesting. While you may enjoy many foods cold, the entire point of refrigeration is, or at least was, to *make foods last*. (See "Food Preservation I" on page 42 and "Food Preservation II" on page 49 for food preservation activities.) The low temperature slows the action of any bacteria in or on foods so that you have days or weeks to eat up the food before it goes bad. The darkness and moisture in the refrigerator are actually conditions that bacteria like for growth. Fortunately, though, the anti-growth effect of the low temperature wins out, and refrigeration is very effective in lengthening the life of fresh foods.

Streptococcus thermophilus

Lactobacillus bulgaricus

The movement of heat is predictable. *Heat always moves toward areas that are cooler.* In other words, the yogurt eventually reaches the same temperature as the room. It's a law, and you can't beat it. But you can slow the process way down. Your warmer does this in two ways. First, the presence of the hot water bottles give you lots of heat to start with (the more warm water you use, the longer the warmth will last). Second, the walls of the cooler are made of a good insulator that holds that heat well. The heat does escape, but only slowly.

You can buy yogurt makers that use electricity to add heat to the milk constantly. They work just as well, but your system *conserved* the heat already in the water and milk. Energy conservation is a good thing! If you insulate your house well and wear the right clothing, you won't need to use so much costly electricity or natural gas to keep comfortable.

13. Sprout Jungle

Vocabulary

Dormant: alive but inactive.
Germination: the process of sprouting.
Photosynthesis: the making of food in green plants. Plants change light energy from the sun into food energy, a form both plants and animals can use for life. Photosynthesis is the beginning of most food chains.
Embryo: an early stage of development. In seeds, the plant-to-be is an embryo.

Cotyledon: a leaf of the bean embryo. Bean seeds have two cotyledons, and they are the first leaves of the sprout. The food supply used during germination is stored here.

What's Going On?

The germination of seeds, sprouting, is triggered by the right combination of moisture, temperature, and, sometimes, light. If seeds were to sprout in the ground right before a snowy winter, the sprout would die and be wasted. So, where winter weather is harsh, seeds stay dormant until spring when conditions trigger fast growth.

Most seeds, mung beans included, contain enough food to give the young plant several days of energy before it must begin to make its own food by photosynthesis. That's why it's possible to grow sprouts in the dark. People tend to like the taste of the whitish, dark-grown mung bean sprouts better. The sprouts become green once moved to the light and can live for a long time. Sprouts kept in complete darkness begin to die from lack of energy after a week or so.

There are no animals that are powered directly by sunlight. Only plants turn sunlight energy into the form of energy stored in food. Plants themselves, after producing food by photosynthesis, use that food energy to grow and to make seeds for reproduction. Animals, including humans, feed on plants and other animals. So life forms all depend on sunlight for survival with plants playing the crucial role at the bottom of the food chain. Honor thy plants! Photosynthesis rocks!

When the weight of a sprout is tracked, the biggest gain in a day occurs when the seeds soak in water at the beginning. They more than double their original weight. Imagine if the same thing happened to you. If you weighed 100 pounds (45 kg), when you went swimming (for 8 hours), you would come out at about 250 pounds (113 kg)! And your swimsuit would look like one of those split-open seed coats. What a sight! Additional weight gain in the seeds comes from more water uptake. When the seeds are in light and begin to photosynthesize, then they also gain weight from storing the food they make.

Because sprouts are small and skinny, they have a lot of surface area for their weight. This allows them to cook quickly. It also results in quick wilting if left sitting out. If a nice, plump sprout is left out for a few hours, its cells lose water and sag, making the sprout lose its crispness.

14. Herb Garden

Vocabulary

Food chain: the series of living things that begins with a plant, the food producer. The producer is eaten by an animal, which may be eaten by another animal, and so on. At the end of the chain are the decomposers such as worms, insects, and bacteria that take apart the bodies of all organisms after they die and return the nutrients to the soil.

Germination: the sprouting of a seed.

Thinning: removing some of the plants to make room for the strongest ones.

Harden off: adjusting plants to outdoor conditions by gradually increasing their time outside.

Stomates: pores on leaves through which oxygen and carbon dioxide enter and exit.

Photosynthesis: the series of chemical reactions in plant cells that result in the making of food molecules from carbon dioxide and water. The reaction is powered by sunlight.

Cellular respiration: the series of chemical reactions that take apart food molecules and release energy for cell use. Both plants and animals carry out cell respiration.

Oxygen gas: the substance that plants and animals must have in order to get energy from their food. Its formula is O_2. Air is about 20% oxygen.

Carbon dioxide gas: The gas used by plants in photosynthesis. It is also produced in both plants and animals by cellular respiration. Its formula is CO_2.

Toxic: poisonous.

What's Going On?

If your plant-parenting experience was a good one, you've been getting a supply of fresh greens. Plants are the first link in food chains because they are able to use the energy of the sun to build food molecules in the process called photosynthesis. No animals have this special ability. Photosynthesis rearranges the atoms in water and carbon dioxide to make sugar, a source of energy. Animal cells make lots of different molecules but can only do so if the basic supply of energy comes from the outside.

While you are harvesting your herbs, you may not realize that the plants are also improving air quality by absorbing some harmful gases and by releasing oxygen. Photosynthesis releases oxygen as a waste product through pores in the leaves called stomates. Some gases that are toxic to people are absorbed by plants as well, reducing indoor air pollution. Unfriendly gases get into the air in your house from paints, insulation, furniture, carpeting, tobacco smoke, cleaning products, inks and glues, dry-cleaned clothing, and plastics. If your home is tightly built, these toxins may not be able to escape. The gases that are absorbed are taken up by the plants through the stomates or they are taken apart by bacteria that live in the roots.

Check the library for good books to learn more about growing herbs. It is an art as well as a science to grow beautiful and healthful herb gardens.

15. Kitchen Compost

Vocabulary

Microbe: a living thing so small that individuals can only be seen with a microscope.

Chemical reaction: a change in matter when atoms rearrange to make at least one new substance. The decomposition of dead plants is a series of chemical reactions.

Decomposition: the chemical breakdown of big molecules into simpler, smaller molecules. The decomposition of dead organisms recycles the materials needed for life.

Decomposers: organisms that feed on things that have died. Decomposers include microbes such as bacteria, fungi, and protozoans and other small organisms such as worms and insects.

Compost or humus: the dark, organic results of decomposition.

Organic: once living.

What's Going On?

Soil is complicated stuff. It is a blend of eroded rock (sand, silt, and clay), water, living critters, and humus. It helps plants grow by holding their roots and by providing the water and minerals they need. Soil quality is higher when it has plenty of humus.

Composting turns plant material into humus by chemically breaking down the materials that made up the plants. In soils high in clay, humus lightens the texture of the soil, letting air and water get in and making it easier for roots to grow. In sandy soil, the humus helps hold onto water and nutrients that would otherwise drain off.

The soil in your bin contains a microbe zoo of many kinds of bacteria, fungi, and protozoans along with other organisms such as worms and insects. With plenty of water and air, these decomposers live well off the food scraps. Their digestive enzymes cut big molecules into smaller molecules and they use the energy that is released to live and multiply. The waste products of their activity form the dark, rich humus or compost.

Decay is fastest when food, oxygen, and water conditions are good for decomposer growth.

- Too much water cuts off their oxygen. Too little water prevents chemical reactions needed for growth.
- Up to a point, warmth speeds up growth, but they will die if it gets too hot.
- Turning increases oxygen supply, promoting growth.
- Some scraps are more easily decomposed than others.
- Decomposition is faster when surface area is increased. Little chunks of foods are digested faster than big chunks and chopped peelings are digested faster than regular peelings. Any additional surface area is like setting more places at the table; the microbes get to the food more easily.

Recycling plant material in your bin or in a larger compost pile outdoors conserves resources. Some communities compost leaves and yard clippings on a large scale and offer the compost back to residents. If yard and kitchen scraps go into a landfill, their soil-replenishing value is wasted. Conditions there are *not* good for microbe growth so decomposition is extremely slow. The material in landfills is compressed so that very little water and air get to the scraps. By keeping plant wastes out of landfills, the limited landfill space lasts longer for the trash we can't recycle so easily.

16. Balance for Small Weights

Vocabulary

Unit: the dimension of a measurement such as pounds, days, inches, or grams. Seconds are units of time. Inches and meters are units of distance.

Center of gravity: a point around which the weight of an object is evenly distributed.

Plumb: vertical; up and down.

Plumb line: a tool that indicates vertical. It is made from a string with an attached weight.

A balance: a weighing tool that has two arms on opposite sides of a pivot point.

A scale: a weighing tool that has a spring or spring-like mechanism that responds to weight. The weight of an object is usually shown on a dial.

What's Going On?

You would think that you would need an expensive, highly technical tool to measure anything with great accuracy, right? Not always! There is one simple tool that always works *perfectly* because it's driven by the very consistent force of gravity. That tool is

Inspections and Dissections / Water in Ketchup?

the plumb line. When an object hangs freely, its center of gravity always comes to rest at a point directly below the point from which it hangs. The bottom of it points to the center of the earth. Bricklayers and wallpaper hangers use this tool, a string with a weight on the end, to find vertical. They need to have their wall or their paper run straight up and down. As long as there's no earthquake going on, a plumb line indicates vertical every time.

It's not obvious right away, but your balance is just like a plumb line. The pin is the hanging point. You make the balance so that the baskets are level when the center of gravity of the straw-plus-baskets is directly below the pin. Adding weight to one side moves the center of gravity of the balance toward that side and the straw tilts to one side. Only by putting the same weight on both sides will the center of gravity move back to the center and put the tool back into the balanced position.

That's how the balance works but what about the units you used with your balance? Units tell you the size of a measurement. Pounds and ounces in the English system or kilograms and grams in the metric system are the units used for many foods.* These are standard units that everyone understands. If someone says, "Here's a pound of butter," you know what to expect. If someone said, "Here's 156,000 mung bean-weights of butter," that would sound a bit odd. But units such as lentil-weights, mung bean-weights, or bead-weights are perfectly fine to use. They are not very standard, but you can handle being unusual now and then, right? Actually, if what you want is a comparison of one weight to another (soaked beans to dry beans or dry ketchup to wet ketchup), the units cancel out anyway. Then it makes no difference

*For a comparison of mass and weight, see page 210 of the Appendix.

whatsoever that you used an unusual unit. Do realize that mung beans (or beads or lentils) are not all identical, so your accuracy is not very good if you use just a couple of them. By chance you might choose unusually light or heavy beans. But when you use quite a few, the accuracy is excellent. The odds are that within the bunch, the number of light beans is about equal to the number of heavy ones and the average of all is really average.

17. Water in Ketchup?

Vocabulary

Estimate: an educated guess.
Dehydration: water loss.
Percent: divided by 100; for example 50% = 50 ÷ 100 = 0.5.
Source of error: a step in a procedure that leads to imperfect results. Every experiment has them.

What's Going On?

Let's get right to the results. Data collected on major ketchup brands show they are about 64% water. That's almost two-thirds water! The water content of living things ranges from about 5% in some dormant seeds to about 95% in jellyfish. Most living things fall into the upper half of this range. Water brings nourishment to cells and carries away wastes. It is required for the many chemical reactions of life. Tomatoes have even more water than ketchup. Some of the water is removed to make ketchup thick.

How does 64% compare to your results and your estimate? It's fine if your estimate was way off, by the way. You used related knowledge such as what kinds of things flow to make a reasonable estimate, but only a good test can be counted on. The whole point of putting effort into scientific investigations is that you can't know certain things any other way.

Scientific methods are not *perfect*, either. So even when you try to do things right, it's important to realize where the weaknesses are in a procedure and to decide how they might affect the results. Imperfections in a procedure are called *sources of error*. You had sources of error. The calculations assumed that *all* the water and *only* the water left the ketchup while it was in the oven. Actually, that ketchup leather could have had a bit of water left in it, especially in any area that was thicker than others. And, remember the aroma of the ketchup when it was drying? If you can smell something it means that particles of that substance have gone through the air. Since water is odorless, some other particles must have evaporated too. So what do you do with these realizations? One, you make note of the observations. Two, you consider the sources of error when you tell your level of confidence in your conclusion. Doing the best possible work, the actual value of the water content could be a little higher or a little lower than what you reported.

18. Seed Survival

Vocabulary

Ovary: part of the base of a flower. It contains the developing seeds.
Fruit: the ripened ovary.
Seed: the reproductive structure that contains embryo and food supply inside a protective coat.
Embryo: the baby plant with the beginnings of roots, stem, and leaves.
Seed coat: the covering that provides seed protection and controls water entry.
Chemical change: a rearrangement of atoms so that at least one new substance is formed.
Proteins: large molecules that form many cell structures and perform many functions in living things.
Denatured protein: a protein molecule that has been changed (damaged) in shape and function.
Propagation: reproduction; starting new plants.
Genes: hereditary material.
Sexual reproduction: reproduction where genes from two parents are combined.
Asexual reproduction: reproduction where the new individual has the same genes as its one parent.
A clone: a living thing that results from asexual reproduction. The offspring has the same genes as its parent. Plants that come from cuttings or from bulbs are clones of the parent.

What's Going On?

It's impressive how much of our food supply comes directly from seeds. Of course roots, stems, leaves, flowers, and fruit also provide food. But seeds tend to have a very concentrated source of nourishment since they don't also have to provide structure for the mature plant. Corn, rice, and/or wheat, all seeds, provide the bulk of the diet in many parts of the world.

You may have been surprised at how many seeds survived the tortures of the experiment. Tolerance of extremes varies with the type of plant and where the plant comes from. But seeds are designed to live through tough conditions such as winter, so they are much more hardy than the plants that grow from them. If a bean *plant* were to freeze, the plentiful water in its cells would expand and crystallize, tearing the cell membranes and cell walls. Death of the plant is certain. The greater resistance of seeds to cold damage relates to their low water content. Less than 15% water, the cell structure is not broken up when the seed freezes. Freezing slows or stops chemical reactions, but it doesn't destroy the substances needed for later growth. Some seeds such as peach seeds actually need to experience winter's cold before they will germinate. Seeds are far more resistant than growing plants to boiling and microwaving, but these treatments are more deadly than freezing. Substances that are needed for normal growth in the cells are destroyed when there is too much heat. Boiling cooks (denatures) the proteins used to carry out chemical changes needed for growth. Microwaving can do the same but only the water in the seeds is heated directly by microwaves. The rest of the seed is heated indirectly by the heated water. With both microwaving and boiling, the seeds can handle small doses, but the longer exposure times are deadly. The microwave dose that the seeds get depends on the total mass of the seeds as well as how long the microwave runs. With the same microwave time, bigger seeds have a better chance of survival. In boiling water, the seed mass is less important. The reason for this is that microwave energy is all directly focused into the seeds. If the energy is spread out over large seeds, it will be less intense than if it is spread out over small seeds. In boiling water, seeds absorb heat by their contact with the 212°F (100°C) water.

Seeds aren't the only method of propagation in plants. Roots and stems can also produce new plant shoots. The new plants formed this way are clones of the parent plant with the same genetic makeup. Seeds, however, are usually a product of sexual reproduction. Pollen from other plant individuals delivers sperm cells that fertilize eggs in the ovary of a flower. The fertilized eggs mature into seeds and so have a mix of genes from two parents.

19. The Case of the Telltale Fingerprint

Vocabulary

Physical evidence: clues related to material things such as fingerprints, blood, hair and fiber samples, and handwriting.
Latent print: a fingerprint of skin oils left on an object.
Forensics: crime detection science; the analysis of any physical evidence from firearms to fingerprints to DNA.

What's Going On?

Fingerprints give us a reliable way to identify people since every finger has a one-of-a-kind pattern of ridges. Even identical twins have different fingerprints. Glands in the skin continuously produce sweat and oil, and these substances are left behind on the objects a person touches. Being colorless, latent prints are almost invisible until they are developed with carbon or some other chemical. Police at a crime scene will "dust for prints" to find and collect fingerprint evidence. Typically, they use a fine carbon powder and apply it with a soft brush. While most features of a person such as bone structure and hair color change with age, the pattern of fingerprints stays the same for life.

Having fingerprints on file assists in solving crimes. Britain's crime investigation department, Scotland Yard, began using a system of fingerprint identification in 1901. The science has grown continuously since then. The FBI, the Federal Bureau of Investigation of the United States, began its file over 20 years later, but now it has the largest collection of fingerprints in the world. Current computing technology has greatly improved the speed and quality of print identification. Prints are added to the file when anyone is arrested.

Fingerprints may be used in other ways as well. Sometimes they are used to identify not criminals, but victims of accidents or violence. Parents often have children fingerprinted to assist police in case of a future incident. Government workers and people entering military service are fingerprinted as are many other employees. Some office security systems use fingerprints to restrict who is allowed to enter.

The FBI classifies prints into eight categories: four kinds of loops, two kinds of whorls, and two kinds of arches. About 65% of prints are loops, 30% are whorls, and 5% are arches.

20. The Incredible Egg

Vocabulary

Yolk: the yellow ball in the egg that contains water, fat, and protein to feed a growing chick.

Yolk membrane: the thin covering that holds the yolk together.

Thin and thick albumen: the clear liquid and the jell in an egg that contain water and protein to feed a growing chick.

Chalazae: the two twisted albumen cords that hold the yolk in the middle of the egg.

Blastodisc: the structure on the surface of the yoke that contains the egg cell. This is the part that grows into a chick embryo if fertilized and kept warm.

Shell membranes: the two coverings that surround the egg inside the shell and keep bacteria out.

Calcium carbonate: the mineral that makes up 98% of the egg shell. Bones, teeth, chalk, and limestone are also mostly calcium carbonate.

Oxygen: the portion of the air that plants and animals must take in to live. Air is about 20% oxygen. Its formula is O_2.

Carbon dioxide gas: The gas produced by the reaction of vinegar or other acid with calcium carbonate. Its formula is CO_2.

Semipermeable membrane: A thin covering that will let certain substances pass through.

Osmosis: the movement of a fluid across a membrane as a result of concentration differences between one side and the other.

Evaporation: the change from the liquid state to the gas state, as in drying.

What's Going On?
Here are the parts of the egg to be labeled:

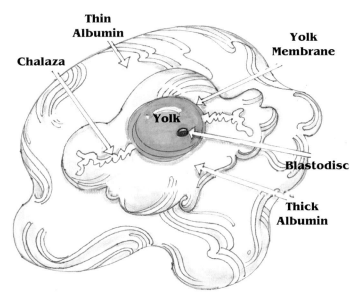

The blastodisc contains a single egg cell, which, in most eggs from the grocery store, is not fertilized. Fertilization occurs early in the egg-making process. If the egg cell is fertilized and kept at the right temperature, it will grow into an embryo and develop into a chick in about three weeks. Notice that the word "egg" gets used for both the tiny single cell that becomes the chick and for the total package of that cell plus the yolk, albumen, and shell. The yolk and the albumen provide all the food and water the growing chick needs. Oxygen reaches the chick by traveling from the air through the shell and the membrane and dissolving into the albumen. The chalazae extend from the yolk into the albumen toward each end of the egg. They hold the yolk cushioned within the albumen. The two shell membranes surround the albumen. They are stuck together except at the air space at the wide end of the egg. The shell membranes keep bacteria out of the egg and slow the loss of water by evaporation. The shell is made of a hard mineral called calcium carbonate. It has thousands of tiny pores that allow the air to slowly pass through. Water in the gas state also escapes out these pores. The shell, the yolk, and the albumen are all a little denser than water, so they sink. The shell membranes are slightly less dense than water. They just barely float.

When an egg is soaked in vinegar, a chemical reaction occurs between the vinegar and the calcium carbonate. As a result, new substances form including carbon dioxide gas that bubbles on the surface of the egg. When a bit of shell is put into vinegar, it sinks at first, being more dense than the vinegar. As little bubbles of CO_2 form on it, the density of the shell-plus-bubbles becomes less than the vinegar, and it rises to the top. If bubbles rub off on the way up or pop at the surface, the piece of shell drops back down and the process repeats. A bubble is like a life jacket for the shell piece. Life jacket on, it floats. Life jacket off, it sinks. Popcorn kernels in a

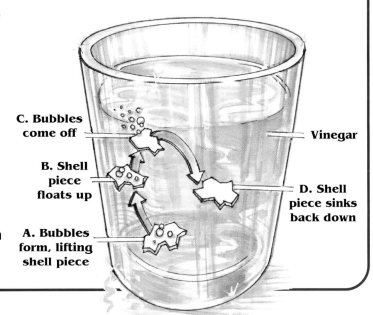

carbonated drink will do the same thing but the shell action is especially cool because the shell is part of the source of the bubbles.

For the whole egg soaking in vinegar, the entire shell is gradually removed. Yet the shell-less egg is actually bigger than it was before. It increases in weight about 40%. This is because water in the vinegar travels through the membrane by osmosis into the egg. The membrane stretches to hold the additional liquid. In the corn syrup, which has a high sugar concentration, water travels out of the egg through the membrane. The membrane loses so much water that it sags. When returned to water, water moves back into the egg, stretching the membrane once again. In each case, water moved from an area of high concentration to an area where it was less concentrated.

Once the egg sits out in the air, water evaporates through the membrane, shrinking the egg slowly. How quickly this happens depends on how humid (moist) the air is. Drying happens with an egg in its shell also, but it loses water much more slowly. Can you give two reasons why eggs need shells?

21. Horoscope Wisdom

Vocabulary

Data: information collected through direct observation or measurement.
Trial: one try in an experiment. In this study, the response of each person to each horoscope version is a trial.
Experimental group: the group of trials that are the focus of your interest. In this case, the trials with the proper horoscopes form your experimental group.
Control group: the group of trials needed for comparison in order to evaluate the experimental group. In this case, the control group is the set of trials with the incorrect signs for the horoscopes.
Variables: the factors or conditions that affect the results of an experiment. There must be only one variable in which the experimental and control groups differ.
Source of error: a flaw in the design of an experiment that can lead to imperfect results.
Order effect: a possible source of error when being first or elsewhere in a sequence favors one group of trials over another.
Psychology: the study of thinking, motivation, belief, emotion, and behavior.
Astronomy: the science of the stars, planets, and the entire universe.
Astrology: the study and practice of predicting human traits and the future based on the positions of heavenly bodies.

What's Going On?

Most testers find that the average rating for the scrambled horoscopes is close to the average rating for the proper horoscopes. There is no big difference. If horoscopes are meaningful, it should be significantly *higher* for the proper horoscopes. If your results were similar, you have to conclude that the horoscopes you tested do not especially fit the people of the proper sign. The information about equally fits any bunch of people.

Of course, your results could vary. You might want to argue with the preceding statement. And if you have real data, you have earned the right to argue! Arguments in science often go on for years until errors are found, better testing methods develop, and/or the vast majority of evidence leans in one direction. Just be sure you have closely examined the design of your experiment for sources of error. So far, serious experimenters have not found evidence that horoscopes provide information special to people born at certain times. Even though people commonly think that horoscopes are accurate, the sense of accuracy comes from other factors. Sometimes information that seems very personal, such as "an important relationship needs your attention," could actually apply to almost everyone.

You might wonder why we had to test two horoscope versions. Why didn't we just ask folks how well their real horoscope fit and stop there? Good question! You need something to compare to. People's beliefs make it hard for them to judge some situations fairly. The way this experiment was designed, we did not focus on their judgment of the horoscope fit alone. We looked to see if peoples' sense of horoscope accuracy depended on the zodiac sign. That way, the degree of people's faith or lack of faith in horoscopes affected both versions the same.

By the way, scientists have to be very clear thinkers and this is one reason we write a clear statement of the purpose of an experiment. Then we can keep checking back to it to make sure our procedure really fulfills the purpose.

If there is no experimental evidence that horoscopes are meaningful, why *do* so many people read and use them? For one thing, there is a great deal of incorrect and misleading information that people pass along through books, classes, web sites, and word of mouth. Errors in most subjects eventually get discovered and corrected. On this subject though, some people resist revising their thinking. Something about the human mind keeps us attached to the ideas that are the most appealing. Human psychology is the science that looks at how we think, learn, and behave. It's a fascinating area for your further study!

Whether it's psychology, astronomy, or chemistry, the whole idea behind science is finding and acting on what is really true. Astrology (horoscopes are a part of astrology) began as a science. The Babylonians used the stars to tell the proper timing for planting and harvesting crops. The seasons are directly connected to the positions of the sun and the stars, so this was effective. The Babylonians divided the path of the sun into twelve constellations of the zodiac and developed the system of hours, minutes, and seconds, all based on the numbers twelve and sixty.* Astrology, evolved, however, into belief that the

*"A Scientist Looks at Astrology" by Eric Carlson, in *Skeptically Speaking*, Volume 1, Issue 1, pages 3–6.

positions of the stars and the planets at the time of a person's birth influence personality and luck. For these things, no connections are known. Astro*nomy* carried on the science of sky observations while astro*logy* became a pseudo-science. *Pseudo* means false or fake.

The zodiac itself is a concept used in astronomy. It is the area of the sky in a band or ring around the earth following the apparent path of the sun during a year. It is 16 degrees wide, and the moon and planets travel within this zone almost entirely. Thirteen areas that cross the center of the band are named by the constellations they contain. These include the twelve names used by astrologers as the signs of the zodiac: Aries, Taurus, Gemini, Cancer, Leo, Virgo, Libra, Scorpius, Sagittarius, Capricornus, Aquarius, and Pisces, plus Ophiuchus, the serpent holder.

The dates assigned to the astrological signs of the zodiac are when the sun is supposed to be traveling across that section. The funny thing is, because the earth wobbles on its axis, the positions of the constellations have shifted since the time those dates were assigned. Astrologers use dates drawn from so long ago that they are way off. For example, *astronomers* observe that the sun moves across Taurus the bull from May 14 to June 19, yet the dates *astrologers* still use for the sign of Taurus the bull are April 21 to May 21.

22. Color Matters

Vocabulary

Data: information collected through direct observation or measurement.
Variables: the factors or conditions that affect the results of an experiment. There must be only one variable in which the experimental and control groups differ. We wanted color to be the only variable to differ in this test.
Consumers: the people who buy and use products.
Trial: one try in an experiment. In this study, the response of each person is one trial.
Sample size: how many people or items are tested. The larger the sample size in a test, the more likely that the results hold true for similar people or items that were not tested.
Experimental group and control group: the two groups of trials that you compare. In many cases, including this one, it doesn't matter which group is called the experimental group and which is called the control. In this experiment, the trials for each color provided the control group for the other color.
Source of error: a flaw in the design of an experiment that can lead to imperfect results.
Order effect: a possible source of error when being first or elsewhere in a sequence favors one group of trials over another group.
Blind test: when a subject does not know information that could lead to prejudice. Taste tests are usually blind tests with products labeled something like "A" and "B" so that only the actual product is compared.
Perception: what you *think* you get from your senses; the brain's interpretation of sensory input.
Natural selection: the shaping of traits in a population. The most useful traits are passed on more than others because the individuals that have them are able to have more babies.

Quantitative: using numbers from measurements or counts.
Qualitative: described in words, without measurements or counts.

What's Going On?

Your flavor ratings for A and B form the core of the experiment. If the average flavor ratings of the two versions matched or were close, you must conclude that either color had no effect or that the two colors you used had the same effect on flavor perception. If the average ratings of A and B were a half a point apart or more, you have evidence that color did affect flavor perception.

It seems that food appearance is a component of flavor perception. You would think that people could separate flavor from looks, but we don't, at least not completely. The brain has its own way of combining sensory input into flavor. Cooks everywhere use this fact to enhance their cooking success. It's called presentation. A cherry pie looks good when it has a shiny crust. In fancy restaurants, the chefs make each course a work of art: from the choice of the plates, to the contrasting food colors, and the swirls of sauce. Food manufacturers set up focus groups to test possible new products. The focus groups are made up of consumers who are given a product to try and asked zillions of questions about it. They are asked not just about taste but about all aspects of appearance including texture, shape, and color. If the taste is good but the color or shape is not the best, then the taste could be even better!

While the brain pathways may not be clear on how this happens, it surely is an advantage to survival to distinguish good and bad colors for food. Imagine monkeys looking for fruit. As fruit ripens and later rots, it changes colors several times. The fruit is most nutritious when ripe. Like kids at an Easter egg hunt, some monkeys will be faster than others at gathering the best fruit in the area. Those individuals will thrive and their genes will eventually be passed on to more babies. This is how natural selection works. The poor monkeys wasting time and energy with fruit that is too young or too old will miss out in more ways than one.

One last point. It would be simpler to ask subjects which of the two products they preferred, rather than do ratings. Why didn't we set up the test that way? *Qualitative* data like that would provide information. But the amount of preference gets lost. Does the subject like the product a lot more or just a bit more? You can't tell. You collected *quantitative* data in this experiment to get the most information and to stay objective. The numbers tell the story and can be compared more fairly.

23. Memory Mystery

Vocabulary

Data: information collected through direct observation or measurement.
Hypothesis: a tentative explanation. Further testing is needed to find out whether it is correct or needs to be changed.
Variables: the factors or conditions that affect the results of an experiment. There must be only one variable in which the experimental and control groups differ.
Trial: one try in an experiment. In this study, the response of each person is one trial.
Sample size: how many people or items are tested. The larger the sample size in a test, the more likely that the results hold true for similar people or items that were not tested.

Experimental group: the group of trials that you are most interested in. In this case, the trials where objects are touched form the experimental group.

Control group: the group of trials needed for comparison in order to evaluate the experimental group. In this case, the control group is the set of trials where the objects are only looked at.

Source of error: a flaw in the design of an experiment that can lead to imperfect results.

Order effect: a possible source of error when being first or elsewhere in a sequence favors one group of trials over another group.

Quantitative: uses numbers from measurements or counts.

Qualitative: described in words, without measurements or counts.

Auditory: related to hearing.

Kinesthetic: related to the sense of body position and movement.

Neuron: nerve cell.

What's Going On?

How memories are stored in the brain is not well understood. *You* are the current expert on the effect of touching objects to remember them. In one group of sixteen eighth-graders, 50% remembered more objects when just looking at them, 31% percent had the same scores whether they touched the objects or not, and 19% scored better when touching the objects as well as looking at them. These results have not been repeated by other experimenters, the sample size was small, and the subjects were all about the same age in the same school, so we can't expect them to represent the general population. How did they compare to your results?

Interviews with these eighth-graders were interesting. People who had higher scores with touching thought the sensations from cold metal or rough surfaces gave them something extra to mentally grab onto when recalling the objects. Some who did better just looking felt the act of touching distracted them from their work of getting the objects into their memory. It seems that different people have different modes of memory making that are strongest. This matches educational research findings that some people learn best from what they see (visual learners), others learn best from what they hear (auditory learners), and others learn best from what they physically do (kinesthetic learners).

Whatever your original hypothesis, realize that there is nothing bad about having a

hypothesis proven to be incorrect. Productive scientists may have many hypotheses proven to be wrong. When you learn a hypothesis is wrong, you form a new hypothesis and keep going. When you learn one is right, you form a hypothesis about the next question and keep going. Knowledge grows either way. If you are too concerned with your hypothesis being right, it can hurt your ability to analyze data fairly.

To understand how the brain actually stores memories, the brain itself must be studied. This is done in medical research facilities around the world. We'd like to know how brain structures and brain cell action are related to memory functions. Studying patients with brain injuries and recording brain activity with electrodes and scans have contributed to our knowledge. Recent brain research has been assisted by a brain imaging technology called functional magnetic resonance imaging, or functional MRI. Researchers can see which structures of the brain are active while a person does memory tasks. Knowledge about the brain is growing.

Here are a few facts about the brain and memory:

- Your brain weighs about 3 pounds (1.4 kg) and contains about 100 billion nerve cells (neurons).
- Neurons receive and send messages as electrical impulses that travel over 200 miles (320 km) per hour.
- Each neuron can "connect" with thousands of other neurons with the help of chemical messages that jump the gaps between them. It may be that these connections are enhanced when memories are formed.
- Repetition, such as repeating the name of a person you have just met, helps move memory from short-term to long-term storage.
- Several areas of the brain work together for memory function.

24. Snack Foods
What Controls Your Taste?

Vocabulary

Data: information collected through direct observation or measurement.

Variables: the factors or conditions that affect the results of an experiment. There must be only one variable in which the experimental and control groups differ.

Consumers: the people who buy and use products.

Trial: one try in an experiment. In this study, the response of each person is one trial.

Sample size: how many people or items are tested. The larger the sample size in a test, the more likely that the results hold true for similar people or items that were not tested.

Experimental group and control group: the two groups of trials that you compare. In this case and many others, it doesn't matter which group is called the experimental group and which is called

the control. The group of trials for each brand serve as the control for the other brand.

Source of error: a flaw in the design of an experiment that can lead to imperfect results.

Order effect: a possible source of error when being first or elsewhere in a sequence favors one group of trials over another group.

Blind test: when a subject does not know information that could lead to prejudice. Usually taste tests are blind tests with products labeled something like "A" and "B" so that only actual taste is compared.

Quantitative: uses numbers from measurements or counts.

Qualitative: described in words, without measurements or counts.

What's Going On?

If the average ratings for the two products were very close, such as a few tenths apart or less, then you have evidence that the images of the brands did not influence your subjects. If, however, the average ratings for the two brands differed by five-tenths or more, then you have evidence that the brand image did affect the taste experience.

Most experimenters who do this kind of study find out that there is prejudice around brand names. This varies with the products compared and the people doing the test. Eight-year-olds and eighty-year-olds may be less tuned in to brand image than thirteen-year-olds. We are prejudiced when we like or dislike something based on things that don't really matter. Advertisers look at this kind of data and use it. They know that product names, package colors, and advertising campaigns all cause reactions in consumers that, *along* with the qualities of the product, determine what we like and buy. Brand images can color our judgment.

Product manufacturers predict the success of a new product before it ever hits the market by testing it on selected subjects. They invest lots of money in market research to avoid big losses that could occur if a product in full production were to fail in sales. Most of these tests predict consumer interest accurately, yet, error is always possible. The folks at sales giant Coca-Cola were caught by surprise in 1985 when their new recipe for Coke called "New Coke" was released. They had tested and interviewed nearly 200,000 people to be sure the new formula and image would be a success. Their research cost $4 million. But they miscalculated the effects of some very loyal consumers who were angry about the switch. The old Coke had been an American tradition for 99 years. The public rejected New Coke and demanded the old formula back. Interestingly, the activists often failed to identify or prefer old Coke in blind taste tests.* Yet, the public reaction was so strong that in less than three months, Coca Cola brought the old recipe back under the name "Coke Classic" and New Coke was phased out. Indeed, food and drink preferences go way beyond actual taste.

One last point: it would be simpler to ask subjects which of the two products they preferred rather than do ratings. Why didn't we set up the test that way? *Qualitative* data like that would provide information. But the amount of preference gets lost. Does the subject like the product a lot more or just a bit more? You can't tell. You collected *quantitative* data in this experiment to get the most information and to stay objective. The numbers tell the story and can be compared more fairly.

God, What a Blunder: The New Coke Story, Michael Bastedo and Angela Davis, December 17, 1993.

INDEXES AND APPENDIX

Life Science Vocabulary and Concept Index

CONCEPT / ACTIVITY NUMBER

Acid 7, 12
Active site 8
Albumin 20
Aroma 2, 8, 17
Asexual reproduction 18
Astrology 21
Astronomy 21
Auditory 23
Bacteria 6, 7, 12, 15
Balance 5, 16
Binocular vision 4
Blastodisc 20
Blind spot 3
Blind test 2, 22, 24
Blood vessels 3
Botany 13, 14, 18
Brain 1, 2, 3, 4, 5, 8, 22, 23
Brand prejudice 24
Butter making 11
Calcium carbonate 20
Calories 9
Carbohydrate 9, 11, 12
Carbon dioxide gas 14, 20
Cell membrane 7
Cell respiration 14
Center of gravity 5, 16
Chalazae 20
Chemical change or reaction; chemistry 6, 8, 9, 11, 14, 15, 18, 20
Churning butter 11
Classification of fingerprints 19
Clone 18
Colloid 11
Compost 15
Conservation 12, 15
Consumers 22, 24
Control group 2, 7, 10, 18, 21, 22, 23, 24
Coordination 5
Cotyledon 13
Crime solving 19
Dairy products 11
Data 10, 21, 22, 23, 24
Decomposition, decay, and decomposers 6, 7, 15

CONCEPT / ACTIVITY NUMBER

Dehydration 6, 10, 17
Denatured protein 8, 12, 18
Density 11, 20
Depth perception 4
Diffusion 7, 20
Dormant 13
Double-blind test 2
Dying of thirst drinking salt water 7
Egg, structure and function 20
Embryo 10, 13
Energy 9
Energy conservation 12
Enzyme 8
Evaporation 6, 20
Experiment design 2, 4, 6, 7, 8, 10, 11, 13, 15, 17, 18, 21, 22, 23, 24
Experimental group 2, 10, 21, 22, 23, 24
Eye-hand coordination 1
Fair test 2, 22
Fat 9, 11
FBI 19
Fingerprints and latent fingerprints 19
Flavor 2, 8, 22
Flower 18
Food chain 13, 14, 15
Food preservation 6, 7
Forensics 19
Fruit 6, 18
Fungi 6, 7
Germination 10, 13, 14, 18
Graphing 2, 4, 8, 10, 18, 21, 22, 23, 24
Hardening off 14
Horoscopes 21
Humus 15
Hypothesis 23
Insulation; insulate; insulators 12
Iodine 9
Kinesthetic sense 1, 5, 23
Lactose 12

CONCEPT / ACTIVITY NUMBER

Lactose intolerance 12
Magnetic resonance imaging 23
Market research 24
Memory 23
Microbes 6, 7, 15
Milk 11, 12
Mirror writing 1
Mold 6
Monocular vision 4
Motion sickness 5
Motor impulses 1
Natural selection 22
Negative test and false negative test 9
Neuron 5, 23
Observations 7, 13, 14
Odor fatigue or odor adaptation 8
Oil 9
Optic nerve 3
Optimum 10
Order effect 8, 21, 22, 23, 24
Organic 15
Osmosis 7, 20
Ovary 18
Oxidation 6
Oxygen 6, 14, 15, 20
Percent, calculation of 17
Perception 3, 4, 5, 22
Photosynthesis 13, 14
Physical change 11
Physical evidence in a crime 19
Plumb line; plumb 16
Positive test and false positive test 9
Preservative 7
Proprioception 1, 5
Protein 8, 11, 12, 18
Psychology 21, 22, 23, 24
Qualitative 21, 22, 23, 24
Quantitative 21, 22, 23, 24
Reception 3, 5

CONCEPT / ACTIVITY NUMBER

Record-keeping 6
Recycle 15
Refrigeration 7, 12
Retina 3
Salt, the action of 7
Sample size 22, 23, 24
Scotland Yard 19
Seed coat 10, 13, 18
Seeds 10, 13, 18
Semicircular canals 5
Semipermeable membrane 20
Sensory impulses 1
Sexual reproduction 18
Shell membranes 20
Smell 2, 8
Soil composition 15
Source of error 17, 21, 22, 23, 24
Spores 7
Sprouts 13
Starch 9, 10
States of matter 10
Sterile; sterilize 7
Stomates 14
Substrate 8
Surface area 6, 13, 17
Taste 2, 8, 22
Touch 5, 23
Toxic 14
Trial 2, 21, 22, 23, 24
Unit of measurement 16
Vacuole 8
Variable 11, 12, 15, 21, 22, 23, 24
Vinegar, the action of 7
Vision 1, 3, 4, 5, 22
Vision therapy 4
Vitamin C 6
Volume 11
Water content of living things 10, 17, 18
Whipped cream 11
Worms 15
Yogurt 12
Yolk 20
Yolk membrane 20
Zodiac 21

National Science Education Standards Matrix

NATIONAL SCIENCE EDUCATION STANDARDS
OF THE NATIONAL RESEARCH COUNCIL, 1996, GRADES 5–8

SCIENCE AS INQUIRY STANDARD (QUOTED)
Abilities Necessary to Do Scientific Inquiry
Identify questions that can be answered through scientific investigation
Design and conduct a scientific investigation
Use appropriate tools and techniques to gather, analyze, and interpret data
Develop descriptions, explanations, predictions, and models using evidence
Think critically and logically to make the relationships between evidence and explanations
Recognize and analyze alternative explanations and predictions
Communicate scientific procedures and explanations
Use mathematics in all aspects of scientific inquiry

LIFE SCIENCE CONTENT STANDARD (ABBREVIATED)
Structure and Function in Living Systems
Living systems demonstrate the complementary nature of structure and function.
All organisms are composed of cells. Most organisms are single cells; others are multicellular.
Cells take in nutrients to provide energy for their work and to make needed materials.
Each type of cell, tissue, and organ has a structure and set of functions that serve the organism as a whole.
Humans have interacting systems for digestion, respiration, reproduction, circulation, excretion, movement control and coordination, and disease protection.
Disease is a breakdown in structures or functions of an organism.
Reproduction and Heredity
Some organisms reproduce asexually. Other organisms reproduce sexually.
In many species including plant species, females produce eggs, and males produce sperm. Their offspring are never identical to either parent.
Heredity is the passage, from one generation to another, of instructions for the traits of an organism.

ACTIVITY																							
1	2	3	4	5	6	7	8	9	10	11	12	13	14	15	16	17	18	19	20	21	22	23	24
X	X	X	X	X	X	X	X		X	X	X	X	X	X		X	X	X	X	X	X	X	X
	X	X	X	X	X	X	X		X	X	X	X	X	X		X	X			X	X	X	X
X	X	X	X	X	X	X	X	X	X	X	X	X	X	X	X	X	X	X		X	X	X	X
X	X	X	X	X	X	X	X	X	X	X	X	X	X	X		X	X	X	X	X	X	X	X
X	X	X	X	X	X	X	X	X	X	X	X	X	X	X		X	X	X		X	X	X	X
X	X	X	X	X	X	X	X	X	X	X	X	X	X	X		X	X			X	X	X	X
X	X	X	X	X	X	X	X		X	X	X	X	X	X		X	X			X	X	X	X
	X	X	X		X		X		X	X	X		X	X		X	X			X	X	X	X
	X	X	X	X					X			X	X			X		X					
		X			X	X	X			X	X							X					
				X	X					X	X	X				X		X					
		X		X	X	X	X						X			X		X					
X	X	X	X	X			X														X	X	X
	X		X	X																			
																	X						
																	X						

LIFE SCIENCE CONTENT STANDARD (ABBREVIATED)
Reproduction and Heredity
Hereditary information is contained in genes, located in the chromosomes of each cell.
The characteristics of an organism are a combination of traits, some inherited and others from interactions with the environment.
Regulation and Behavior
All organisms must be able to obtain and use resources, grow, reproduce, and maintain stable internal conditions.
Regulation of an organism's internal environment involves sensing the internal environment and changing physiological activities.
A behavioral response requires coordination and communication at many levels, including cells, organ systems, and whole organisms.
An organism's behavior evolves through adaptation to its environment.
Populations and Ecosystems
All populations living together and the physical factors with which they interact compose an ecosystem.
Populations of organisms can be categorized by the function they serve in an ecosystem: producers, consumers, decomposers.
Sunlight energy is transferred by producers into chemical energy through photosynthesis and then passed from organism to organism in food webs.
The number of organisms an ecosystem supports depends on resources including abiotic factors such as light, water, and soil composition.
Diversity and Adaptions of Organisms
Millions of species of animals, plants, and microorganisms are alive today and have similar physical structures, chemical processes, and ancestry.
Evolution accounts for species diversity. Species acquire unique characteristics over many generations through the selection of natural variations in populations.
Extinction of a species occurs when the environment changes and the adaptive characteristics of a species are insufficient to allow its survival.

National Science Education Standards Matrix

203

ACTIVITY

1	2	3	4	5	6	7	8	9	10	11	12	13	14	15	16	17	18	19	20	21	22	23	24

					×	×					×	×		×			×						
×	×			×								×		×									
×	×	×	×	×																			
×	×		×	×			×																

													×										
												×	×	×									
												×	×	×									

														×									
																					×		

Graphing Appendix

A calculator is recommended.

Graphs are information pictures. The idea is to show data in a way that is understandable at a glance.

The results of most experiments are shown in a **bar graph** or a **line graph**. A third type, a **circle graph**, is used when you want to show the relationship of parts to a whole.

In an experiment, you try to find the effect of one variable, called the *manipulated variable* on another variable called the *responding variable*. In "Kitchen Compost," for example, you might test the effect of a type of food scrap (manipulated variable) on the time it takes the food to decompose (responding variable).

To be able to create a bar or line graph, your responding variable must be quantitative—a measurement or a count, such as time to decompose—so that you can put the results on a number line. This variable goes on the vertical (y) axis so that the reader's eye moves UP when values are large and DOWN when values are small.

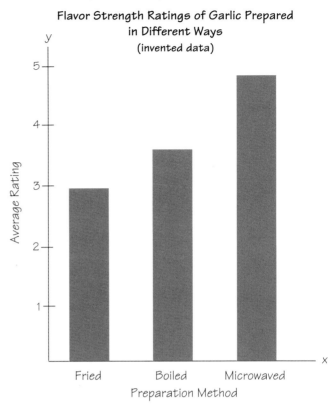

Bar Graph

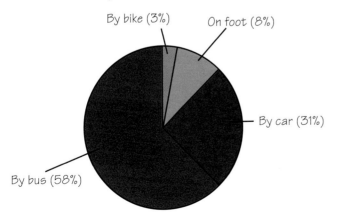

Circle Graph

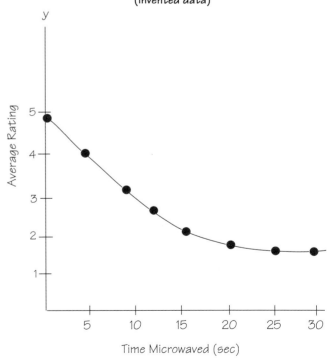

Line Graph

Graphing Appendix

If the data for the manipulated variable is *not* a measurement or a count—for example, it's carrot peels vs. cucumber peels—then you make a bar graph.

If the data from the manipulated variable can be put on a number line—for example, time in the blender—then you make a line graph.

How to Make a Bar Graph

A bar graph is made when just one variable can be put on a number line.

Example: Let's say your results for "Kitchen Compost" were as shown in this table:

TIME TO DECOMPOSE (DAYS)		
CARROT PEELS	CUCUMBER PEELS	EGGPLANT PEELS
8	3	13

This data would be made into a bar graph. The responding variable, time to decompose, can be put on a number line, but the manipulated variable, type of vegetable, cannot. There will be three bars.

1. Get a sheet of graph paper*, a ruler, and a well-sharpened pencil. Your graph will be made to fill the sheet. Choose which way to place the page.

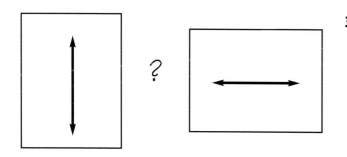

*Graph paper is provided at the end of this book.

2. Use a ruler to draw in the vertical (*y*) and horizontal (*x*) axes. You need to leave enough space outside the lines for labels, so indent them several squares from the edges of the paper. Also end the lines a few squares before you reach the top and right edges of the paper.

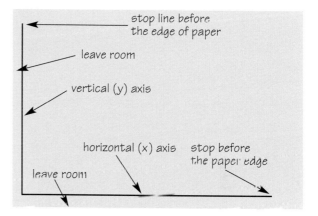

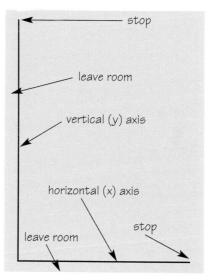

3. On the horizontal (*x*) axis, you are going to mark off equal-width bars with spaces between them. Properly dividing up any graph axis is the hardest job in graphing. Sometimes, you can see instantly how to do it or you can work it out with trial and error. The following steps will direct you if you think about their logic as you go:

a. Count the number of graph paper boxes available on the axis.
b. Divide the number of boxes (use a calculator if you like) by the number of bars you have, and round your answer *down* to the nearest whole number.
c. Split that number into two parts, one for the space before the bar, and one for the bar itself. In other words, for the answer 10, you might decide to make each space 4 boxes wide and each bar 6 boxes wide, or 5 boxes for spaces and 5 boxes for bars.
d. Mark off the bars and spaces on the horizontal axis. Write in labels centered below the bars in any reasonable order. You might want to order them from tallest to shortest or shortest to tallest.

4. Now you will create the number line on the vertical (y) axis so that your highest value will be toward the top of the paper. Realize that your data values are NOT the values you put along the axis unless it just happens to work out that way. You will label the axis next to the graph paper *lines*, not the *spaces*. You might see instantly how to spread out the number line or you can work it out with trial and error. Or, you can follow these steps:
   a. Pick a round number just above the highest value of the data. This number will end the number line segment you will fit to the vertical axis.
   b. Divide the number of boxes you can use on the graph paper (count them) by the number you just selected.
   c. Round your answer *down* to a nice, round value such as 0.1, 0.2, 0.5, 1,

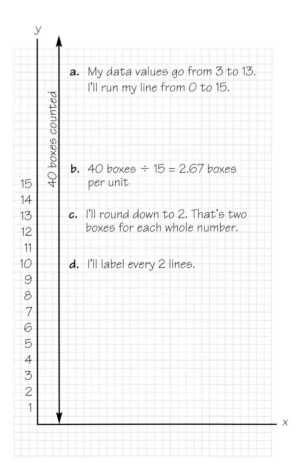

a. 27 boxes counted
b. 27 boxes ÷ 3 bars = 9 boxes per bar (including space between) No rounding needed since it came out to a whole number.
c. I'll do 4 boxes for each space and 5 boxes for each bar.

space bar space bar space bar

Cucumber peels   Carrot peels   Eggplant peels

d. I'll arrange the bars from shortest to tallest.

40 boxes counted

a. My data values go from 3 to 13. I'll run my line from 0 to 15.
b. 40 boxes ÷ 15 = 2.67 boxes per unit
c. I'll round down to 2. That's two boxes for each whole number.
d. I'll label every 2 lines.

Graphing Appendix

2, 5, 10, 20, 50, or 100. This will be the number of boxes for each whole unit on the number line.

d. Write in the number value for every two, five, or ten lines.

5. Add labels for each variable outside each axis. The vertical scale will have units such as meters or seconds in parentheses next to the label, unless it is a count.
6. Use the ruler to help you estimate the appropriate heights of the bars, draw them, and shade them in.
7. Add a descriptive title at the top of the page. Hey, that is looking good!

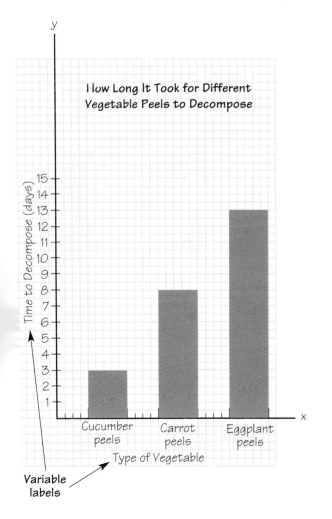

How to Make a Line Graph

A line graph is done when data from BOTH variables can be put on number lines.

Example: Let's say you did an experiment in "Color Matters" where subjects rated the sweetness of 7-Up with different amounts of red food coloring, and your results look like this:

Amount of Color (drops per liter)	Average Sweetness Rating (1–10 scale)
0	5.5
2	5.7
4	6.0
6	6.5
8	6.4

This data would be made into a line graph because both variables, the Amount of Coloring and the Average Sweetness Rating, can be put onto number lines. The sweetness was the outcome, the responding variable, so it goes on the vertical axis.

1. Get a sheet of graph paper, a ruler, and a well-sharpened pencil. Your graph will be made to fill the sheet. Choose which way to place the page.

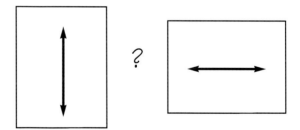

2. Use the ruler to draw in the vertical (*y*) and horizontal (*x*) axes. You must leave enough space outside the lines for labels, so indent them several squares from the edges of the paper. Also end the lines a few squares before reaching the top and right edges of the page.
3. As with bar graphs, the responding variable (the outcome) goes on the vertical (*y*) axis. The other variable is

called the manipulated variable, and it goes . . . you guessed it, on the horizontal (*x*) axis. The manipulated variable either changes naturally, such as time, *or* is set by the experimenter, such as amount of food coloring.

4. Number each axis in even intervals so that the smallest value lands close to an axis line and the largest value lands near the opposite end of the paper. This step is the hardest part of making the graph. You might see instantly how to do it, or you might work it out with trial and error. The following directions work well especially if you think as you go:

For one axis at a time,

   a. Pick round numbers just below the lowest data value and just above the highest to be the beginning and end of the number line for the axis.
   b. Subtract the lower value from the higher one.
   c. Count the number of boxes you can use on the graph paper. Divide that number by your answer to b.
   d. Round your answer *down* to a nice, round value such as 0.1, 0.2, 0.5, 1, 2, 5, 10, 20, 50, or 100. This will be the number of boxes on the graph paper per whole unit on the number line.
   e. Write in the value of every two, three, four, five, or ten lines.

5. Add the variable labels and units outside each axis.

6. Plot the points. For each point this requires estimating where the data values fall on the number line of each axis and then following imaginary lines to the right from the vertical axis and up

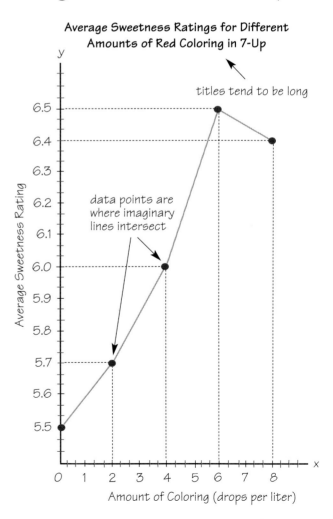

For vertical axis (y):
   a. The data run from 5.5 to 6.5. Those are pretty round numbers as is.
   b. 6.5 − 5.5 = 1
   c. There are 38 boxes I can use.
       38 boxes ÷ 1 = 38 boxes per unit
   d. I'll round this to 30. That's 30 boxes from 5.5 to 6.5.
   e. I'll label every third line.

For horizontal axis (x):
   a. The data run from 0 to 8. They are already round numbers.
   b. No need to subtract since they start at zero.
   c. 27 boxes ÷ 8 = 3.375 boxes per unit
   d. I'll round this down to 3. That's 3 boxes for every whole number.
   e. I'll label every third line.

Graphing Appendix

from the horizontal axis until they intersect.

7. When all points are plotted, connect the points with a pencil line or draw a best fit curve.*
8. Add a descriptive title to the top of the page, and you are done!

*Best Fit Curve

When you have quite a few data points, it is better to draw what is called a best fit curve than to connect the points. What you do is stand back and look at the trend your graph shows. Then draw the line to be smooth and continuous. (It may be curved or straight.) It's OK to miss some or even all of the points as long as you follow the trend they seem to show. If the line misses some points, it should miss about as many above the line as below the line. The best fit curve assumes that there was some experimental error. It is your estimate of the true results of your study.

The line graph at the beginning of this appendix was drawn with a best fit curve.

How to Make a Circle Graph

A circle graph is done when you want to show the parts of a whole.

Example: Let's say that you did "Water in Ketchup?" and found that your brand of ketchup was 68% water. You could show this with a circle graph. You need a compass, or a drinking glass, and a protractor.

1. Draw a circle with a compass or by tracing something circular.
2. Turn percents into degrees. The entire way around the circle is 360 degrees.

Example: 68% of 360 is 0.68 × 360 = 244.8, or about 245 degrees. The nonwater part is what's left, or 360 − 245 = 115.

3. Use the protractor to draw wedges in the circle measuring the proper number of degrees. Label each wedge.
4. Add a title to your graph.

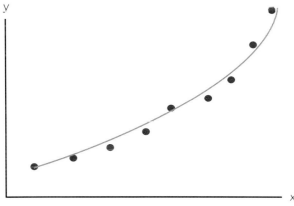
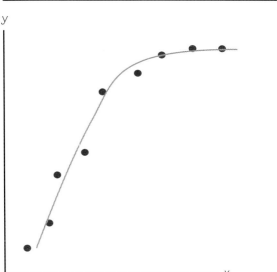

Examples of Best Fit Curves

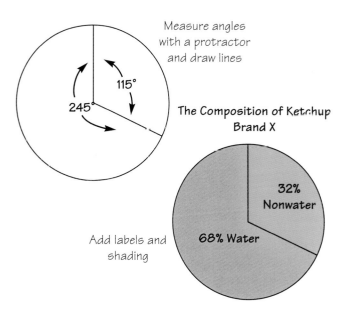

Key to Metric Abbreviations

Units of Length

mm = millimeter (or 0.001 meter)
cm = centimeter (or 0.01 meter)
m = meter
km = kilometer (or 1,000 meters)

Units of Mass

g = gram (or 0.001 kilogram)
kg = kilogram (or 1,000 grams)

Units of Weight

N = newton

Units of Volume

mL = milliliter (or 0.001 liter)
L = liter

Units of Temperature

°C = degrees Celsius

Unmuddling Mass and Weight

	Mass	**Weight**
Is it a way to measure matter?	Yes	Yes
Is it a force?	No	Yes
Does it have a direction?	No	Yes, downward
Does it depend on gravity?	No	Yes
Would yours change if you moved to the moon or a different planet?	No	Yes—The gravity there would be different. The more gravity, the more weight.
What does it depend on?	The amount and type of material in the object	The mass of the object and the gravity where you are
What are the English (or Imperial) units for it?	Slugs Pounds and ounces mass	Pounds and ounces
What are the metric units for it?	Kilograms and grams	Newtons

Key to Metric Abbreviations

You can see why these terms get so confused. "Pounds" and "ounces" get used as both units of mass and weight. If words crystallize concepts, the units for mass and weight give us no help. Science books tend to use metric units where it is clear that mass and weight are different quantities. Ten U.S. quarters or a medium apple each have a mass of about 100 grams and weigh about 1 newton on Earth. On the moon, the quarters and the apple each still have a mass of 100 grams, but they each weigh only one-sixth of a newton. They are lighter because gravity is weaker on the moon. They are pulled toward the moon with one-sixth of the force they experience on earth. Want to lose weight? Go to the moon! You won't look any different, but you'll feel lighter. A girl with a mass of 27 kilograms weighs 60 pounds on Earth but only 10 pounds on the moon. Her mass is 27 kilograms no matter where she goes.

Grams (Mass) vs. Ounces (Weight) on Earth

On earth, 28.4 grams of mass weigh 1 ounce (one-sixteenth of a pound).

To find the mass in grams of a certain weight in ounces (on Earth), multiply by 28.4.

Example: 4 ounces × 28.4 grams per ounce = 114 grams

To find the weight in ounces (on Earth) of a certain mass in grams, multiply by 0.0352.

Example: 100 grams × 0.0352 ounces per gram = 3.52 ounces